The Black Rock That Built America

The Black Rock That Built America

A Tribute to the Anthracite Coal Miners

Gerald L. McKerns C.C.

Contents

DEDICATION

(a)

In loving memory of my daughter, Karen Ann, who after three short days of life was taken away by an angel. Because of her bloodline, the good Lord had for her a special assignment. Her task, although difficult, would be noble. No heavenly sounds from harps or trumpets would be heard. The only music to reach her ears would be the rhythmic beat of a pick clanging against a rock face. You will find her in an anthracite coal mine shadowing a miner as he toils for his daily bread deep in the bowels of the earth. If fate would have it that the roof fell in around him, she would stay by his side to give him hope and comfort until he is freed from his prison and again sees the light of day. What a blessing it is to be the father of an angel.

Introduction

The starlit night is giving way to the first light of day as the sun begins to inch its way over the horizon. Dew on the blades of grass sparkle as the sun's warm rays reach its moisture. It's a new day, a new beginning.

The sound of alarm clocks going off would be the first sound many of us hear on this early morning. Time to get up, slumbers over, it's time to start our day. Out the door we go. There's no time to waste. Next, we stop to get some coffee and pick up the morning newspaper. This is how many of us start our day. By country lanes, crowded freeways, or mass transit, off to our jobs we go.

The sun shone so brightly on September 11, 2001. Who would have thought that it would have been so darkened by smoke and dust on that tragic day? A day history will never forget. The sun would set that evening on a saddened nation. There would be an ache in the hearts of us all.

We are told we should move ahead. The history of the United States reveals many crisis's throughout the years. After each, we stood united and pushed ahead. We never stood still and always became a greater nation afterwards.

Principals set down by our forefathers in the form of a constitution are only a dream for some other countries. Freedom to voice our opinion, to choose our political leaders, our places of worship and where we live and work are truly a gift.

This country, as we know it today, with its economic prosperity and military might, didn't just happen. It took years of hard work by the many that came before us. We are blessed with vast natural resources, but our greatest natural resource is our people. The people who took a wilderness and with their sweat and blood would have it evolve into what we have with our constitution as its cornerstone and our people as its building blocks a great nation did emerge.

Many of our children aren't aware of the hardships their grandfathers and great-grandfathers endured to sculpt this nation into its grandeur.

Every day has a yesterday and every today has a tomorrow. As each tomorrow arrives, each yesterday gets further into the past. The stories our great-grandfathers handed down by word of mouth get fewer with each passing generation, some day they will be no more.

To appreciate where we are today it is essential that we know where we came from.

Growing up in the hard *coal region* of northeastern Pennsylvania, I wasn't aware, nor was I taught, the importance this small area of our country was in propelling this vast wilderness into a global giant.

It all began in the early nineteenth century for those looking to escape the poverty and unemployment of Europe. The anthracite coal region of northeastern Pennsylvania appeared to be the land of opportunity. They would come by the thousands drawn by the promise of steady work in the mines. The first wave of immigrants rolled across the Atlantic from Wales, England, and Germany in the 1820s. In the 1850s, most of the immigrants were Irish. By the 1870s, there was a steady influx of Poles, Slovaks, Russians, and Lithuanians. By about 1910, many more were coming from Italy. At one point, twenty different languages were spoken in the coal region. Most of the immigrants were hired by the mining companies and did the grueling and dangerous work of extracting coal from the earth to earn their place in America. (1)

The wages were low and the price was high. Loss of life would be the price many would pay for their work in the mines. By the year 2000, the price in human life in the anthracite coal region, reached 31,115 men and boys. (2) Some of these lads were as young as ten years of age.

Loss of life was so commonplace, in the hard coal region, that it often didn't make the front page of the local newspaper.

I can remember one such incident that happened when I was growing up in Mahanoy City, in Schuylkill County. It was a warm summer day. I was a lad of about nine years of age when I heard one long blast from the whistle. One long blast was a signal that there was some type of accident or fire on the outskirts of town and volunteer help was needed.

Emergency vehicles headed north and so did I. About a mile up the road, I saw a dozen or so cars and pick-up trucks pulled haphazardly off to the side of the road. A dirt road veered off to the right. I could hear men's voices coming from that direction so I followed the sounds of the voices a short distance up the road. As I got closer, I could hear the excitement in their tones. When I saw an ambulance parked, I knew I was near the activity. I did not want to get too close as to be in the way so I climbed a tree. Once I got up about ten-feet, I was able to see what was going on. To my right there was a coal mine going into the side of a mountain. What I saw was a human chain of men frantically passing rocks from one to another out of the mine. The vigorous pace that these men were working told me that there must be a miner trapped down below. I glanced to my left and I saw a house. It was an old beat up house. Its back door was opened. On the doorstep sat a young boy about the age of six. He sat watching the men. His elbows rested on his knees and his chin rested in the palms of his hands. The blank look on his face was telling me that it was his father trapped in that mine.

These men were working against the clock but never stopped or even slowed. One man, I remember in particular, had a white T-shirt on that was now full of sweat and grime. The stripe running down the side of his pants told me that he was a mailman. He wore gloves, but the fingers were worn thru. I was close enough to see that his fingers were bleeding. His hands probably weren't as calloused as the others were.

After about an hour and a half I saw a man coming out of the mine. The look on his face was troubled. He talked softly to a few of the men at the mouth of the mine. Then a deafening silence came over the mountain. Not even the birds were chirping. It was over. The miner's fate was known.

I turned my head to the boy. He raised himself up to his feet, took a deep breath, turned, and went into the house.

At that time, I got down out of the tree and started home. It had now turned into a recovery operation, not a rescue.

The next day I got a call from our parish priest asking me if I was available to serve as an alter boy the next day for a funeral. I was. As the service began, I looked to the back of the church and coming up the isle to the front pew was the boy from the doorway and his mother. I didn't know the service was for the miner.

From the church, we went to the cemetery for the burial. I looked at the boy comforting his mother and I couldn't help to think of the many children who had lost their fathers and the many mothers who were widowed because the family breadwinner was a coal miner.

Just the other day I watched men remove dirt that covered a miner who was in a hole in the ground and today men will put that miner in a hole in the ground and cover him with dirt.

This scene took place over thirty-one-thousand times in the anthracite coal region of Pennsylvania. So many men, so many families and all the miners wanted to do is provide for their families the only way they knew how.

Chapter One

ON THE EIGHTH DAY HE RESTED

The sun is taking its daily journey lazily toward the hills off in the distance leaving behind in its wake a magnificent display of pastels interwoven amongst the clouds. Soon its brightness will be completely swallowed up by the hills and night will be upon us.

Even though darkness has set in around us, our homes will be well illuminated by electric lighting that turn night into day with the flick of a switch.

This country uses over 3.7 trillion kilowatts of electricity per year and the demand is growing. More than half of the electric generating plants in this country burn coal as its source of energy. Almost 91% of the coal mined in the United States is used for this purpose. If we were to replace the amount of coal used to produce electricity with gas or oil the price of electricity would sky rocket.

Besides electricity, coal's heat and by-products are used in a variety of industries. Separated ingredients of coal, such as methanol and ethylene, are used in making plastics, tar, synthetic fibers, fertilizers, and medicine.

It takes huge amounts of coal to fuel this country's hunger for energy and huge amounts of coal we have. In the United States we take over a billion tons of coal from the earth each year. Even at that rate of consumption, we have enough coal in the ground to last for hundreds of years. (3) That vast amount of coal below us is like a giant underground storage battery waiting to be tapped in order to continue to light our cities, energize our computers, and turn the wheels of our industries for generations to come.

The evolution of coal in Pennsylvania began over two-hundred-fifty-million years ago. Both anthracite and bituminous were formed approximately the same time. The state was a flat, hot, moist plain covered with steamy swamps, thick with tall trees and wide spread ferns. As the giant plants and ferns of this tropical age died and decayed, they fell to the bottom of swamps where they mixed with the remains of animal life

Bituminous (soft coal)

Anthracite (hard coal) (b)

such as fish, amphibians, reptiles, and insects. The mass eventually formed a spongy brown vegetable matter called peat. Over millions of years, this peat was squeezed down and sealed off from the air by mud, sand, and new debris carried in by the vast shallow inland sea that covered the middle of the continent. The resulting increased pressure combined with heat rising from the earth's core to drive off moisture and gasses transformed the peat, over millions of years, first into lignite and then into bituminous or soft coal.

The relentless, eons long, natural process created in parts of Pennsylvania, as many as one-hundred superimposed beds of coal, reaching a total thickness of thirty-five-hundred-feet and separated by vast sedimentary layers of sandstone, shale, clay, and limestone. (4)

Today as I scan the landscape of the coal region it certainly doesn't look like a flat, hot, moist plain covered by steamy swamps, giant plants and ferns. I see rugged mountains covered with boulders and lush dense forests. A land where, before it was settled, the Indians rarely ventured, they called it *towamensing* (wild place). (5)

14

This is a land where the temperature could vary more than one-hundred degrees from winter to summer.

This is a land where you see mountain streams run swiftly down steep ridges being fed by voluminous amounts of water produced by the melting snows of winter.

This is a land where, in the spring, you see the fresh greenness of the forest saving room for the Mountain Laurel that dot the green forest with its dainty white blossoms.

This is a land where warm summers ensure an abundant supply of fresh blueberries on the mountain tops; a sweet gift from nature.

This is a land where in the fall you see the forest taking on a breathtaking display of colors as the trees ready themselves for their winters rest.

This is a land where even in the dead of winter life can still be seen on the mountainside with the evergreen trees and the green leaves of the Mountain Laurel protruding from the deep snow.

This is the land we call the *coal region.*

Not only did the landscape change over millions of years, but so did the location. When most of the world's coal was forming the area, which would eventually become North America was situated in the tropics, immediately north of the equator. Various proto continents had been drifting toward each other and by the end of the carboniferous period, the earth's crustal movement culminated in a four-way collision, resulting in the plates of the earth's crust welding all the worlds' landmasses into a single super continent. (6)

The land-sea patterns of today have evolved over the course of hundreds of millions of years, during which time continental landmasses drifted, were united by collisions, then torn apart and recombined. These movements show no signs of slackening. So the distribution of sea and land will continue for as long as the planet contains the heat energy required to drive the movements of its crustal plates. (7)

With all these changes that occurred, the Almighty was not yet finished with his job. He wanted to be sure everything would be just right for when it was time for man to arrive on the scene. Would there be enough fish in the sea, game to hunt, water to drink, trees to build shelter and fertile land to grow crops? Everything seemed adequate to meet man's needs. However, with his almighty wisdom he saw a potential problem down the road. What he saw was millions of immigrants leaving their homes in Europe, crossing the Atlantic and arriving at what they would call the New World. They would dock at cities on the east coast; places like Philadelphia, New York, Boston, and Baltimore. Knowing that these cities could not meet the needs of this great multitude of people he had to make a move. Looking down from above he searched for a suitable place to handle the overflow. It should be a place centrally located to these seaports; far enough away, so they could call it their own but close enough so they could get there. The ideal spot would be northeastern Pennsylvania. But would this land meet the needs of so many? If the land could not support so many people, maybe he could doctor it up! And doctor it up he did. The adjustments to this land would be known as the Appalachian Revolution, which consisted of subsequent upheavals of the earth's crust, creating the mountains of the anthracite coal region. With the bituminous coal beds and the additional pressure on them caused by the earth's convections it would change the bituminous coal into a much harder coal with an unusually high carbon content.(8) This coal would be called anthracite.

The anthracite coal region is bound on the north and west by the branches of the Susquehanna River, on the east by the Lehigh River, further north by the Delaware, and to the south by the first ridges of the Blue Mountains, part of Pennsylvania's Appalachian Mountains. The area is a small piece of America; yet it contains three-quarters of the earth's hard coal deposits. (9)

Now the job is done. When the new comers arrive in the area, there will be an abundant supply of hard coal that will be in demand by the consumer, resulting in work so the immigrants could earn a living and feed their families.

What a perfect plan. If when the book of Genesis was recorded, it would have been known what transpired on the other side of the globe, it may have read, "On the seventh day he created the coal region and on the eighth day, while looking down at his masterpiece, he rested."

Chapter Two

BUILDING A NATION

Once again, I find myself driving up Rt. 61 going north from Reading. I have made this journey many times. The dense humid air caused by the hot summer day obscures the Blue Mountain, usually visible on a clear day. As I get closer, its outline against the sky becomes more defined and you now can see its grandeur. From this point, it looks like an enormous barrier. A barrier that separates one world from another and as you reach the other side of this mountain, you are in a different world. You are in Pennsylvania's hard coal region.

My trip today from Reading to Mahanoy City should take me a little over an hour. If the same trip were taken some two-hundred years ago, it

Three generations of miners. (e)

would have taken days. We sure did come a long way. The reason we came so far so fast is because of what took place on the other side of the Blue Mountain many years ago.

America's transformation from a mostly rural nation into the world's greatest industrial power all started in the coal region. As the nation expanded in the nineteenth century, anthracite coal fueled the making of steel, the building of railroads, the operation of factories, and the heating of homes. No other place in America possessed such a natural monopoly as the anthracite coal region. (10)

Before the discovery of anthracite, the area did not attract many settlers. The mountains were rugged and road less and in most of its valleys the soil was hard and rocky, unlike the rich farmland that laid to the south.

The first settlers coming from the south, from well established places such as Reading, Easton, and Bethlehem, entered the region through gaps in the mountain walls chiseled out eons past by rapid running rivers. Settlers from the north entered the area through the thick forests and rolling hills of the Colony of New York. These founding families moved into an area that Pennsylvania's government, in a swindle comparable to the purchase of Manhattan Island, had bought from the tribes of six nations for about two-thousand dollars. The buyers did not know that they had purchased almost all of Pennsylvania's anthracite. (11)

What a bargain; two-thousand dollars in exchange for seventy-five percent of all the earth's hard coal deposits located in this small area of America consisting of less than five-hundred square miles. Having all that coal under foot one-hundred years ago would be comparable today with having seventy-five percent of the world's oil reserves at our disposal. Anthracite coal was king then as oil is today.

History indicates that in 1790 anthracite coal was discovered in Schuylkill County at the southern end of the coalfields in northeastern Pennsylvania. These coalfields extend about fifty miles east and west, one-hundred miles north and south and contain the richest deposits of anthracite coal in the world. (12)

Coal's discovery in the Schuylkill area, as in other areas of the coal region, came about by accident; or we could call it fate. A man by the name of Necho Allen who, it was rumored, decided to spend the night at the base of Broad Mountain after hunting one day in 1790. He lit a fire under a protecting ledge and fell asleep next to it. Sometime in the middle of the night, he was awakened by a bright light and heat. Jumping up in surprise, he discovered, as he said latter, "that the mountain was on fire." When daylight came, he found that what he thought was a rock ledge was actually an outcrop of coal that his campfire had ignited. The fire forecast events to come. (13)

Coal as a source of fuel was not accepted on a large scale until decades later. Wood was plentiful and anthracite was hard to ignite. Some of the local blacksmiths of the area experimented with it and found it preferable over wood charcoal because it produced a much greater heat resulting in the melting of iron much faster.

By the late 1700s, Philadelphia consumed more than a thousand tons of coal annually but most of it was imported from Europe. (14)

On one occasion, the public's resistance to anthracite almost cost the life of a rigorous promoter. In 1812, Colonel George Shoemaker carted nine wagonloads to Philadelphia and talked himself horse trying to persuade homeowners to buy it. He finally sold two wagonloads and gave the rest away. But when it would not ignite, the Colonel was nearly lynched by an irate mob that denounced him as a "swindler and imposter". Fortunately, for Shoemaker's reputation, some of the coal found its way to a rolling mill in Delaware County where, with some experimentation, it was used successfully. (15)

The problem of burning anthracite in a home fireplace was solved when Judge Jesse Fell, of Wilkes-Barre, experimented with anthracite for home heating. Most of his efforts occurred at night because he did not want to arouse the scorn of his neighbors for attempting to burn stone coal in an open grate. He devised an L-shaped grate made of simple iron bars spaced several inches apart, which he inserted in his home fireplace. On it, he placed wood, which he ignited, and then a quantity of coal. A practical man, he did not want to see the coal ignite so he went to bed. When he awakened the next morning, the fire was still giving off a warm, friendly glow. A revolution in home heating had begun.

The judge's discovery caused a great deal of local excitement. If early history is believable, the event stimulated more discussion at church that week than the sermon. As for Jesse Fell, he simply wrote, "Made the experiment of burning the common stone coal in a grate, in a common fireplace, at less expense than burning wood in the common way."(16)

Jacob Cist contributed more than any other single person did to the early promotion of anthracite, and he must be counted among the decisively important forerunners of the American industrial revolution. (17)

Cists' opportunity to promote anthracite came during the War of 1812. The British blockade of American ports, combined with the increasing cost of firewood because of the scarcity of readily accessible supplies near the large eastern cities, provided a chance for hard coal to capture new markets in the midst of America's first fuel crisis. (18)

Cists' principal advertising device was testimonials. Whenever a blacksmith, ironmaster, or other manufacturer experimented successfully with anthracite, the promoter encouraged him to write about it in detail The testimonials were then printed on handbills and circulars and were distributed widely, often by street urchins who ran from place to place for a few pence. Cist also wrote articles for newspapers and magazines extolling the virtues of his product and the best ways of using it. He argued that hard coal burned longer and provided heat that is more regular and that its efficiency as a fuel allowed for an increase in production, and the absence of smoke and fumes improved working conditions and thus benefited the workers as well. Since he was also concerned about pollution, he noted that using anthracite assured a clean environment, an important consideration in a growing industrial city like Philadelphia.

Nor did Cist ignore the potentially vast domestic market. A calculation by James Ronaldson in 1815 estimated that Philadelphia's sixteen thousand households paid out approximately 1.1 million dollars annually to heat their homes. When households were combined with businesses and industries, the city's total fuel bill for wood came to about two million dollars a year. Coal offered an attractive alternative to increasingly expensive hardwoods for fuel and cooking. Here also Cist realized that he would have to educate the public. To do so he designed several stoves for burning anthracite. He had his stoves assembled in the homes of several prominent citizens, in businesses and banks, and even in a jail. These were places where a good many people might be expected to see them. He wrote more newspaper articles explaining that anthracite gave off a constant, regular heat that required little attention and that its clean-burning qualities reduced the chances that the chimney's and stovepipes would catch fire. He also calculated the cost of heating an average-sized room for a year, the quantity of coal required, and the cost-savings benefit when compared with wood.

Cists' promotional methods and similar efforts by others gradually paid off. (19)

Now that there was a potential market for anthracite in the Philadelphia area, the problem of how to get the coal to market existed. The coal region was still much of a wilderness. Its few roads were rocky and impassable when the winter snows came. The answer was the rivers. A canal was built along the Schuylkill River that has its beginnings in Schuylkill County's mountains and flows all the way to Philadelphia.

In 1822, it was reported that 1488 tons of anthracite had been shipped by canal from the Schuylkill region and the industry, as a business, had its beginning. Development was rapid. By 1825, the Schuylkill Canal was completed, providing transportation of anthracite from Pottsville to Philadelphia. The barges of "arks" originally pulled by men using breast bars and long towropes took six-weeks to travel the one-hundred-eight-miles. After towpaths for mules were laid parallel to the canals, the barge sizes were increased from twenty-eight tons to two-hundred tons and the amount of coal transported grew dramatically. (20)

The Blue Mountain Dam and lock, Schuylkill Navigation,
Hamburg, Pennsylvania, 1913.
Courtesy of the Pennsylvania Canal Society Collection, Canal
Museum, Easton, Pa.

(f)

Within a decade, after the canal's opening, the Schuylkill field was the region's most populous and productive. Towns and villages appeared virtually overnight. Port Carbon had a single family in 1829; a year later, it claimed 912 residents. Minersville, New Castle, St. Clair, Tamaqua, and other settlements experienced the same rapid growth. Between 1826 and 1829, the number of buildings in Pottsville increased six fold and its population twenty-seven times and between 1829 and 1844, the number of inhabitants doubled again. (21)

The life of a canaler, however, was anything but romantic. Hours were long, boats moved in all but the worst weather, and there were plenty of hazards. For example, men were sometimes thrown off a badly steered boat and crushed in a lock between the wall and the hull.

On the Schuylkill, the working day began between three and four in the morning. The bows-man prepared breakfast while the driver boy fed, curried, and harnessed the team. The captain's order, "Gear up your team first; then you'll get your breakfast," established the order of priorities. By four, the boat was usually on the move, the way lit by the Nighthawk kerosene lamp on the bow. Stops were infrequent, except for waiting to be "locked through" or perhaps to trade some coal for homemade bread and

Kelly's Lock Muhlenberg Township, near Reading Pa. (g)

pies at nearby homesteads. As the boat approached the locks, the captain announced its coming by sounding a horn or a seashell, called a conch. On an average day, the boat traveled about twenty-five miles before being tied up for the night at about nine. Perhaps the hardest job fell to the driver. He was often a boy no more than nine or ten years old who walked, often barefoot, with the mules throughout the days. On the Schuylkill, there were usually three mules in a team, the lead mule that led and pulled, a middle mule that pulled, and the *shafter*. The towline tended to pull the shafter toward the canal, and it was this animal's job, while being pulled sideways, to keep the other mules from going into the water. The driver's job was to walk behind the mules and keep them in line. He often did this with a whip, not to beat the mules but to give them direction. Some drivers became so proficient with their whips that it was said they could flick flies from the lead mule's ear. (22)

The boatmen were a tough and independent breed not particularly noted for their gentleness. In the early days, it was relatively easy to get a boat, a fact that attracted the footloose and irresponsible types found in abundance in the booming coal region. The hoodlums, drunkards, and ne'er-do-wells tended to give the occupation a doubtful place in the public's eye. Insults were frequently exchanged between passing boats. At locks, fistfights often broke out over which boat was to go through first. (23)

But even the most peace-loving canalers were ready for anything, including piracy. In Philadelphia, a gang known as the Schuylkill Rangers was notorious for attacks on canal boat crews. One night, however, the gang made the mistake of attempting to rob the canal boat called the Rattlesnake. Captain Peter Berger, who had a reputation as a fighter, drew his pistol and shot one of the would-be robbers dead. A magistrate subsequently freed the captain with the remark, "Your pistol did not work well this time. You only killed one man." (24)

The Schuylkill Canal, like the other anthracite canals, opened the region not only to markets for its primary export but also for the outside capital that ultimately made the region an internal colony. (25)

About this time, the industrial revolution was in its embryo stage.

In 1822, Abraham Potts built a new conveyance for hauling coal at his Black Valley Mine in Schuylkill County. The contraption attracted a lot of attention because anything that made it possible to bypass the terrible roads in the Schuylkill wilderness guaranteed to arouse considerable interest. Eventually, news of Potts innovation reached Philadelphia and a group of Schuylkill Navigation Company officers came into the region on a canal boat to investigate. When they arrived, they saw thirteen loaded coal cars standing on wooden rails that extended from the head of the canal at Mill Creek to the mine a half a mile away. The affable Potts quickly offered a demonstration. He hitched up a single horse to the led car, ignored some skeptical wise cracks from the group and had the animal rather easily draw the load over the rails. He then informed the men that within a decade they would see coal from the region travel to Philadelphia entirely by rail. As they stepped onto the boat for the return trip, the company officers promised to lock up Potts as a lunatic should he ever venture into the City of Brotherly Love. (26)

The first rail lines were actually part of the canal system. The rails were feeder systems for carrying coal directly from the mines to the canals. Built in the 1820s and 1830s they were simple extensions of the rails outside the mines. The railways were made of timber and secured on notched crossties bound with iron strapping spiked to the rails. Horses or mules pulled the cars. As more mines opened, lateral lines were built that picked up cars from the feeder lines of collieries along the rout. The lateral roads, chartered by the state as public highways, charged tolls, usually a cent and a half per ton per mile, for use of the track. They were built as cooperative ventures by groups of mine owners who shared in stocks and profits or by independent promoters who arranged with mine owners to have secure outlets to the canals. (27)

In the late 1820s, Americans were reading about a steam-powered locomotive being used successfully in England. It was said that this incredible iron horse could pull heavy loads over rails. It did not make much sense to most folk on this side of the Atlantic. We already had plenty of mules that were low maintenance. All you had to do was feed them some hay and they'll pull all the coal you wanted. Mules were also much cheaper. All you had to do to get a new one was get two of the right kind, make sure they had a smile on their face and Mother Nature would take care of the assembly.

Some men weren't skeptics. One such man was Frederick List from Reading. List was the editor of the German-American newspaper the *Readinger Adler* (Reading Eagle). He predicted that railroads would play a pivotal role in the development of the anthracite industry, and he gambled his own money on it, buying coal lands along the Little Schuylkill River near present-day Tamaqua and persuading a group of prominent Reading citizens to organize a railroad company that would link these coal purchases with the Schuylkill Canal. When construction costs outran the ability of the Reading investors to raise capital, List turned to Philadelphia, the nation's financial center, and interested Stephen Girard, America's wealthiest man, in the project. With the help of Girard and other Philadelphia financiers, the Little Schuylkill Navigation, Railroad and Coal Company, linking Tamaqua with Port Clinton, was completed in 1831. It was one of the first railroad corporations chartered in Pennsylvania. (28)

The Delaware and Hudson Canal Company purchased the first steam-powered locomotive. In 1828, John B. Jeruis, head of the company, sent Horatio Allen, an ambitious twenty-five year old engineer, to England to size up the possibilities of the new iron horses that Americans were reading about in the newspapers. Greatly impressed by what he saw, Allen placed orders for four iron horses and in the spring of 1829, the first four-wheeled locomotive arrived in New York City. It looked, said one observer, like a giant grasshopper. The driving wheels were of oak wood banded with heavy wrought iron, the front was ornamented with a large, fierce looking face of a lion, and it bore the name of the Stourbridge-Lion.

That summer Allen sent the locomotive by river and canal to Honesdale, Pennsylvania, and in early August, he was ready to test it. Many were worried that the Stourbridge-Lion would break the cracked and warped timber railway that ran to the coal mine at Carbondale, or that it would jump the tracks and plunge into the creek. Allen remarked, "My reply to such apprehension was that it was too late to consider the probability of such occurrences that there was no other course but to have the trial made of the strange animal which had been brought here at such great expense, but that it was not necessary that more than one should be involved in its fate that I would take the first ride alone."

Allen recalled that as he placed his hands on the handle of the throttle valve, he "was undecided whether I would move slowly or with a fair degree of speed; but believing that the road would prove safe, and preferring, if we did go down, to go down handsomely and without any evidence of timidity, I started with considerable velocity, passed the curve over the creek safely, and was soon out of hearing of the cheers of the large assemblage present. At the end of two or three miles, I reversed the valves and returned without incident to the place of starting, having made the first railroad trip by locomotive in the western hemisphere." However, the experiments with the Lion, proved ultimately unsuccessful. The primitive rails were not sufficiently strong enough to carry a heavy locomotive. Nor was the engine constructed well enough to pull laden coal cars, but Allen's ride on the Lion inaugurated a transportation revolution in America. (29)

Stourbridge-Lion (h)

That first train ride in the coal region took America out of the embryo stage of the industrial revolution and triggered the birth of an industrial giant. As with any newborn, it must learn to crawl before it walks and walk before it runs. Run it would, but first it must learn to stand on its own two feet. Iron would be the answer for America's growing pains. The nourishment needed for this young country to grow would be anthracite. A surge in the railroad industry was on the horizon.

Some of the early railroads were:

The Maunch Chunk Railroad—Laid out in 1818, was completed nine years later by the Lehigh Coal and Navigation Company. It was nine miles long.

The Mill Creek Mine, Navigation, and Railroad Company—Incorporated in 1828, a railroad four miles long, built from Port Carbon on the Schuylkill Navigation Canal to St. Clair.

The Norweigen and Mount Carbon Railroad Company—Extending from Mount Carbon to several miles west of Pottsville in 1831 and was six miles long.

The Schuylkill Valley Railroad—Chartered in 1827 and completed in 1831, it ran from Port Carbon to Tuscarora, a distance of ten miles.

The Mine Hill and Schuylkill Haven Railroad—Extended from Schuylkill Haven on the canal to the coal fields north and south of the Broad Mountain. It was completed in April 1831

The Little Schuylkill Railroad—Chartered on April 14, 1828 and completed in 1831, it connected Port Clinton with Tamaqua, a distance of twenty-two miles.

The Beaver Meadow Railroad and Coal Company—Chartered in 1830 and completed in 1836, it extended twenty-six miles from Beaver Meadows, in Luzerne County to the Lehigh Canal at Penn Haven.

The Hazleton Railroad Company—Organized in 1836 after the discovery of coal by John Charles in 1826. It ran from Hazleton to Weatherly, where it connected to the Beaver Meadow Railroad.

The Lehigh and Susquehanna Railroad—Started in 1837 and completed in 1843, a distance of about twenty miles from Wilkes-Barre to White Haven on the Lehigh River.

The Pennsylvania Coal Company—Chartered in 1838 with provisions for both the mining and transportation of coal. In 1848, the company began construction of a gravity railroad forty-seven miles long.

The Philadelphia and Reading Railroad—Chartered on April 4, 1833, to build a railroad from Reading to Philadelphia. It was opened as far as Mount Carbon on January 1, 1842. In

1872, by acquisition of the Mount Carbon Railroad, the P&R reached Pottsville.

The Delaware, Lackawanna, and Western Railroad—Opened to traffic in October, 1851

The Leggett's Gap Railroad—Started in 1851 and the line was opened to traffic in October of that year. In April 1851, the corporate name was changed to The Lackawanna and Western Railroad Company.

The Lehigh Valley Railroad—Built shortly after the middle of the nineteenth century became a great anthracite carrier although it was not a pioneer in the anthracite region.

Some of the smaller railroads of the region are listed below with the dates of the charters or openings:

1832—The Northern Central Railroad
1833—The Danville and Pottsville Railroad
1836—The Catawissa Railroad
1840—The Quakake Railroad
1842—The Mount Carbon and Port Carbon Railroad
1844—The Schuylkill and Susquehanna Railroad
1854—The East Mahanoy Railroad
1854—The Treverton, Mahanoy & Susquehanna Railroad
1857—The Lehigh & Mahanoy Railroad
1861—The Nesquehoning Valley Railroad (30)

A surge in heavy industry was on the up swing. America could not keep up with the demand for iron rails. America's charcoal fired iron industry producing iron in small furnaces, in small quantities, and of poor quality just wasn't acceptable. The newfound railroad companies were forced to turn to England to import our much needed iron rails. It wouldn't be long before our suppliers could not keep up with America's hunger for iron rails. In order to grow we must learn to feed ourselves.

In 1840, the success of the Lehigh Crane and Iron Company in using anthracite to make pig iron gave a new impetus to the industry. Production of pig iron soared as new furnaces were built. There were six anthracite furnaces in the United States, all in Pennsylvania that made iron at the end of 1840. Six years later, there were forty-two such furnaces in Pennsylvania and New Jersey. By 1856, there were one-hundred-twenty-one anthracite furnaces in the country. Ninety-three in Pennsylvania, with many more built in the following years. Pig iron production is estimated to have increased nationally from approximately two-hundred-twenty-thousand tons in 1842 to more than seven-hundred-fifty-thousand tons in 1847. The mid-1840s anthracite was making produced pig iron in Pennsylvania for twelve dollars a ton compared to sixteen dollars a ton for charcoal made pig iron. (31)

The railroads were the major beneficiaries of the anthracite iron industry. However, the growth of one would have been impossible without the other. For many years, the railroads consumed as much as fifty percent of the iron and steel industries total output in the form of rails, locomotives, rail cars, and railroad bridges. But anthracite iron making also spawned a multitude of other related industrial activities in the region. Iron works in the regions' cities and towns turned out boilers, pumps, tools, fencing, pipes, structural iron for bridges and buildings, and a variety of other products. (32)

The rapid advancement of the railroad industry for the main purpose of transporting coal found the Philadelphia and Reading Railroad one of the largest industries in the nation. The Philadelphia and Reading Railroad was briefly the largest industrial corporation in the world just prior to the Civil War. (33)

One of the Reading Company's iron horses that pulled loaded coal cars to the canal dock, circa 1887. Courtesy of the Historical Society of Schuylkill County, Pottsville, Pa.

(i)

Thanks to anthracite, this nation climbed out of its cradle and learned to stand on its own two feet. It wasn't long before America was running with the rest of the world, giving chase. It was anthracite coal that gave this country the kick-start it needed to propel us ahead of our neighbors from across the Atlantic, a position we never relinquished.

Anthracite coal production rose steadily until it peaked in 1917 with the production for that year set at over one-hundred million tons mined. (34) From this point in time, except for the years of World War II, there was a steady decline in the production of anthracite.

A series of developments occurred following the war that steadily undermined anthracite's supremacy as a heating fuel. Technological improvements allowed for the increasing use of oil and later natural gas. Though the initial costs of heating plants were high, the safety, convenience, and cleanliness of these fuels, touted in high-pressure advertising campaigns, appealed to affluent families. The new heating substitutes made their greatest inroads in the densely populated, urbanized northeast, where anthracite had its biggest market. Even bituminous, long scorned as dirty and inefficient by anthracite users, successfully challenged hard coal. Better firing and heating equipment made soft coal attractive in distant markets like New England and the Midwest, where transportation costs pushed the price of anthracite even higher. By the mid-1920s, the cost differential between the two coals in these areas was often as much as seven dollars a ton.

Oil, gas, and electricity for domestic use were the products of technological change of an advanced industrial economy that anthracite originally helped create. From the 1920s on, the increasing adoption of these new fuels brought about anthracite's decline. That decline was amazingly rapid for an industry that had seemed so solidly established. (35)

Today coal isn't king. Many of the rails that had millions of tons of coal pass over them are now quiet. But the fact remains that this country went from a wilderness to an industrial giant because of the industrial coal region of northeastern Pennsylvania.

Some of our nations' landmarks had their beginnings in the hard coal region of Pennsylvania. The Bethlehem Steel Company used anthracite coal from the Lehigh Valley as its source of energy to produce its steel. The mill that sits along the Lehigh River rolled the first wide-flanged structural shapes in America. These shapes ushered in the age of the skyscraper and the Bethlehem plant was a leading supplier of steel beams for many of our nation's monumental structures.

Here are some examples:

>The Golden Gate Bridge—San Francisco, California
>Rockefeller Plaza—New York, New York
>The Ben Franklin Bridge—Philadelphia, Pennsylvania
>The Chrysler Building—New York, New York

Anthracite coal helped sculpt this country into America the beautiful. When we hear the phrase, *God Bless America*, that prayer was answered many times over.

Chapter Three

FINDING A PLACE IN AMERICA

It won't be long now. In less than a half-hour, we should be touching down at the Shannon Airport. We have been in the air for about seven hours. Seven hours seems long, but not long at all when you consider the distance, we traveled. As I look out the window through the misty rain I can see the coastline approaching. While making our decent the breathtaking beauty of the Emerald Isle becomes more evident as we get nearer. What a pleasant sensation to see the land that my great-grandparents sailed from some one-hundred-fifty years ago-Ireland.

My trip, although too short, was most enjoyable and will be remembered forever. While admiring the alluring countryside the day I was to return home, I couldn't help but to think of the thoughts that might have been running through my great-grandparents minds on the day they were to set sail for America. It must have been so hard to leave a land so beautiful with its forty shades of green, a land that is certainly an emerald in God's global patchwork.

My trip home would be in comfort. Read a little, watch a movie and enjoy my thoughts and before you know it, I'm home. The same trip my great-grandparents took wouldn't be as easy.

Imagine, it's the mid-1800s and you are leaving your home for the first time in your entire life knowing there is little chance you will ever return. The ship you must travel on is crowded and it's hard to imagine how any attempt at modesty will be made. You make your way through the dark tunnels marked Steerage and you can't help but to notice the overwhelming stench of urine and vomit. You are told that each day you will receive a small biscuit and one-third quart of water. The water must be used for both cooking and cleaning although it is hardly sufficient for one or the other. Small bunks are nailed to the walls and you wonder how all these people will find a space to sleep tonight. People are pushing and shoving in a whirl of confusion. You ask yourself what you are doing on this ship, leaving the only home you've ever known. It's to late to get off.

The ship has already left port and the captain won't turn back. Your only goal now is to survive.

However, there is hope. They believed that it would be better to take the chance of dying on a ship than it would be to remain in the midst of Ireland's poverty and disease.

The *coffin ships* were the unstable vessels that carried emigrants out of Ireland. They were called coffin ships because many people died on them. Like a coffin, one was trapped in a wooden confinement. Most ships lost as much as one-third of their passengers to disease. Much of the food was rotten but passengers could not refuse what was offered. The starving Irish ate moldy, hard food without complaint and longed for more.

The lack of privacy or bathrooms also created a huge problem. On some ships, there were no toilets. Even on ships where there were one or two water closets for several hundred passengers, people were often denied access to them during a storm. Of course you can't keep your bowels from moving so people were forced to relieve themselves wherever they were, often below in the storage deck in the company of rats. It's hard to imagine traveling for up to ten weeks under these deplorable conditions. Though the conditions on the coffin ships may have been worse than the conditions in Ireland, emigrants had the promise of hope. They truly believed that they might find a better life in America.

Finally, land on the horizon, soon this horrible journey will come to an end; you finally reached America. You are too weak to stand but someone lifts you up. For the first time you breathe in some of that sweet American air, everything comes flowing back to you and you remember why you left. You had a dream for a new and better life. (36)

Like so many before and so many to follow, your exodus will take you to the hard coal region of northeastern Pennsylvania. Here is where your new life will take root. Your final destination will be a small coal-mining patch called St. Nicholas located in West Mahanoy Township, Schuylkill County. This is a village with only two streets, one called the "high road" and the other the "low road".

About fifty families resided in this small coal region community. Upon arrival, it's hard not to notice the dismal, sooty, gray dust that coat these clapboard dwellings. The dust is expelled into the air as it exits the coal breaker that sits down the road. From forty shades of green to forty shades of gray, St. Nicholas is where you will earn your daily bread, raise your family, and obtain your rightful place in America. Not much, but certainly better than the accommodation you had aboard that sailing vessel. That company owned house in St. Nicholas—although gone today, as they weren't built to stand the test of time—will always remain special in the hearts of my family. My grandfather was born there, as was my father, and I am proud to say that I was the last one in my family to call St. Nicholas my birthplace. There were years of tears and tales, laughter and sadness, good times and bad. What a testimony of courage and endurance for people, like my great-grandfather, who would survive the great famine in Ireland, survive the journey of hope aboard the coffin ship, and survive a daily reprieve of death while earning a living deep in a coal mine.

In the hard coal region of northeastern Pennsylvania, tragedy came and it came often. The toll of dead and injured was appalling. Three miners were killed every two days and tens of thousands were seriously injured or maimed. Some of the fatalities were caused by great disasters that claimed scores of lives. Most died in accidents involving one or two men. The day-to-day toll of lives lost went largely unnoticed by the public, but the total over time was grim evidence of the human cost of mining coal. (37)

A typical back yard in one of the fast growing mine "patch" communities. (j)

The tragedy that claimed the most lives was the Avondale Mine Disaster that took one-hundred-ten men and boys on September 6, 1869.

The ventilating furnace setting fire to the woodwork in the shaft caused the catastrophe. The stable boss of the mine discovered the fire about nine in the morning. On reaching the bottom, and discovering the fire, he immediately gave the alarm. In a few minutes, a cloud of smoke followed by a mass of living flame rose through the up cast compartment of the mine. The flames set fire to the breaker. The people on the top of the shaft became paralyzed with terror, not knowing the fate of the miners in the distant chambers of the mine. The news of the accident spread like wild fire and people rushed to the burning mine by the thousands to assist in rescuing the imperiled miners, but they were powerless before the burning elements. Streams of water were turned into the burning mine, but the monstrous volume of lurid flames appeared to bid defiance to the water, and for several hours the fire raged with unabated fury. When at last it had become subdued, a band of volunteers offered to go down into the shaft to rescue the imprisoned men or perish in the attempt. A dog and lamp were lowered down as far as possible and on being withdrawn, the dog was still alive and the lamp still burning. This told the rescue workers that the air quality was safe at that depth.

An exploring party was lowered but was unable to withstand the deadly gases and they soon returned. On being raised to the surface, they were nearly overcome by the fumes.

During the second day, several attempts were made to reach the entombed men, but the accumulating gases prevented any extended search. At midnight the air had become greatly improved and at two in the morning an exploring party came upon two dead bodies but they could not recognize their features because of their blackened and distorted appearance.

The explorers returned to communicate the fact to the people above ground. Preparations were made at once for the decent of several bands of explorers. At half past six, as an exploring party was traversing the east side of the mine, they discovered the whole workforce of the mine lying behind an embankment that they had erected to try to shut off the deadly gases.

Fathers and sons were found clasped in each other's arms. Some of the dead were kneeling, as if in prayer. Some lay on the ground with their faces downward, as if they were trying to extract one last mouthful of air from the floor of the mine. Some were sitting with clasped hands as if they had vowed to die with each other, and some appeared to have fallen while walking. In two hours, sixty dead bodies were sent to the surface, and by noon the last of the unfortunate one-hundred-ten men and boys who had gone down to work there days before, full of life and vigor, only to find themselves dying in a tomb and raised from this tomb to be buried. (38)

Some accidents could not have been avoided; mining was a dangerous job that left people subject to the whim of nature. Other accidents resulted from the lack of supervision or by carelessness, loose rock that should have been removed, the lack of timbers that should have been installed for stability, improved blasting techniques, or failure to follow company regulations.

This was the case in the Wilkes-Barre's Baltimore tunnel explosion in 1919, when seven men carried kegs of black powder on the man-trip cars, even though company regulations forbid it.

The man-trip cars had descended two-hundred-feet when the dynamite exploded roasting men alive, blowing others to bits, and suffocating still more. In all, the explosion killed ninety-two miners.

There were times when the colliery continued to operate in spite of dangerous conditions. In 1896, the miners at the Pittston Twin Shaft Colliery knew the roof was unstable and notified the mine officials, rather than stop working, the company ordered additional men to put in extra props over the weekend, so the miners still could work Monday.

However, in the middle of Saturday night, over one-hundred acres of rock and coal collapsed. Fifty-eight men, who were stabilizing timbers, were buried alive. Rescue operations began immediately but were halted as the roof continued to fall in and the Susquehanna River rose slowly in the gangway. As days passed, all hope of saving the miners disappeared and their bodies were never recovered. (39)

Water also posed a major danger to miners. Shafts or slopes were driven beneath the natural water level, causing downward seepage. Mines were generally graded so that water would accumulate at the bottom of the shaft where the sump was located. From there it was pumped to the surface by powerful steam engines through a timbered compartment of the shaft known as the pump way. At the turn of the century, the most powerful pumping engines extracted twelve-hundred gallons of water per-minute. Seldom, in the anthracite coal region, did the tonnage of water pumped from a mine fall below the tonnage of coal hoisted. In the Lehigh fields there were mines where eight to ten tons of water was pumped for every ton of coal extracted. In active workings, every effort was made to control the amount of water, but abandoned workings at the higher level often filled with water. Until accurate underground survey maps were required, miners unaware of the exact location of such workings would sometimes break through a barrier pillar and would suddenly be swept away as the water broke through with irresistible force. (40)

At Jeansville, a mine patch south of Hazleton, such an accident occurred on February 4, 1891. Two miners were firing near an abandoned slope that had been flooded several years before. Their mine map indicated that they were about sixty-feet from the slope, but they were actually only five-feet away. When their blast went off, it opened the old workings and the water poured in with flood force. The two miners were in the gangway and the water swept them to a place where they could swim to safety. Seventeen other men working nearby were not so fortunate; they drowned in the rush of water. Emergency pumps were installed that raised two-thousand gallons of water per-minute. It was several days before the water level was sufficiently low enough so the rescuers could descend. Then they had to make their way through the flooded and rock clogged workings on a raft. After twenty days, five men were found alive. They had managed to survive by eating part of their clothing and by chewing bark from the timbers. Eventually, six bodies were discovered, eaten by rats almost beyond recognition. (41)

Disasters sometimes occurred because of unexpected geological conditions. A Susquehanna Coal Company miner working near Nanticoke on December 18, 1885, broke into a geological depression, or pocket, filled with water, quicksand and culm. Within minutes, twenty-six men and boys had lost their lives. Though efforts to reach the flooded and clogged part of the mine went on for weeks, the quantity of water and debris was so great that rescue attempts had to be abandoned. (42)

Family life was a cycle of drudgery revolving around the start and stop whistles from the mines. The men, including male children as young as seven or eight, trudged off with their lunch pails into the predawn darkness. The women then turned to the household chores for which they had total responsibility. They did the shopping, cooking, cleaning, laundering, and gardening. They cared for sick children and injured workers. They often used their skills with ingenuity, taking in laundering, sewing, and tailoring to supplement family incomes. Often they had complete control of the family budget.

Women also gathered firewood and scrounged lumps of coal from the culm banks of mine waste. Toward evening, they went out to meet the traveling beer wagon so they would have buckets of beer waiting for the returning men. When the men arrived at the house, they would strip and the wives would pour hot water over them while scrubbing them down. At night, the mattresses and bedrolls were spread out in any available space. Young children usually slept with parents for warmth. The daily cycle ended with exhausted sleep. (43)

Another responsibility, the most terrible of all, fell on the women. When, at an odd hour of the day, a shrill continuous whistle blast announced that an accident had occurred down below, the women rushed to the mine, waiting long agonizing hours while rescue operations proceeded. If luck held, there were only injuries. Too frequently, however, dirt caked, mangled, and burned bodies were brought up to the surface. A company wagon then carted the corpses to the miner's homes, sometimes depositing the body unceremoniously on the kitchen floor. Women then had the responsibility of washing and preparing their man for burial. (44)

In addition, if that wasn't enough, now with no pay coming, they cannot deduct the rent for the company house that you live in and you face eviction. The widow could remain in the house with her children if the financial obligation with the coal company was satisfied.

Women get water from a coal "patch" hydrant that supplied water for twenty-four families.

(k)

One heart braking story about Mrs. Kate Burns, who went to work scrubbing floors and taking in wash in order to support her children, the youngest a boy of eight, after her husband had been killed while employed by the G. B. Markle Company. When her boy reached fourteen, he went to work in the breaker. But after a months work, he brought home only a due bill for $396.00, the amount of accumulated rent on the two room shack in which the family had lived since their father's death. Mrs. Burns, her son and another younger son, worked twelve years to pay off that debt. (45)

The supply of cheap labor was reduced as immigration slowed during World War I and afterward. As a result of worker opposition, the most flagrant abuses of company stores and company housing ended. In the largely urbanized anthracite region, where most people lived in towns of more than twenty-five hundred people, the majority of workers owned or rented their homes. And in the tiny, isolated patch towns, prosperity brought new services such as electricity, municipal water, and public schools.

As the Irish, Slavs, and Italians elbowed their way into local public offices, they confidently used their newfound political power to improve their communities. The rising standard of living helped reduce ethnic conflict and social discrimination. The miners' long struggles for a living wage and the right to raise their families decently appeared finally to pay off. Many of them undoubtedly believed that they were on the verge of realizing the hopes that had first brought them to America. (46)

As America turned to alternate fuels, like gas and oil, it brought hard times to the coal region and as the demand for coal dwindled, it brought more and more lay-offs in the mining industry.

The proximity of thin coal seams to the surface, especially in the rugged Schuylkill field, provided a major temptation to unemployed miners. The seams were not substantial enough to be worked profitably by the big companies, but for a father and son and a few friends willing to sink a "dog hole" and operate with primitive equipment they represented bread on the table. The first bootleggers used the coal for their own fuel or bartered it for goods, but soon pickup trucks were delivering it to eastern Pennsylvania cities where it was sold door-to-door at discount prices. By 1933 an estimated twenty thousand men engaged in bootleg mining in an illegal industry worth thirty-five million dollars. Naturally, the coal companies were enraged by the practice. They brought in state police, dynamited the illegal holes, began surface strip mining of the seams, and demanded that the authorities act against the bootleggers. Nothing worked. One company blasted shut more than one-thousand bootleg mines on its properties one year, only to see several thousand more opened the next. A stripping operator had his steam shovel blown up and a Coal and Iron policeman who threatened a group of bootleggers had his car burned. When he showed up the following day on a horse, the miners shot the horse. Local officials were obviously reluctant to tell a miner that he could not support his family. (47)

By the end of the 1930s, hard coal was finished as a major industry. In 1938, production was down to forty-six million tons. There were still about ninety-seven thousand employees in the industry, but payrolls were fifty-percent shorter than ten years earlier. Prosperity never returned. The long painful dying continued during the following two decades as the nation completed its switch to alternate fuels. Mines closed and breakers shut down; anthracite railroads began their slide toward bankruptcy. The old miners worked on as long as they could, but their sons, for the first time, did not follow their fathers into the mines. By the 1950s, the sight of weary, dirt-encrusted middle-aged men straggling home after a shift in the few working deep mines was already a curiosity. They symbolized an era now past and seemed as utterly obsolete as the industry in which they labored.

In the decades after World War II, the out-migration became a flood. The young left first, followed by heads of families. Eventually the middle-aged and some of the elderly went. The coal region lost a quarter of its people. Some of the mining towns lost fifty percent of their inhabitants; a few disappeared entirely. Unemployment never fell below ten percent, even during the expansive 1950s; usually it was much higher. The coal region had, and still has, the lowest per capita income in Pennsylvania and one of the lowest in the entire northeast. Until the 1960s, unemployment compensation provided the major source of income in the largest cities of Scranton and Wilkes-Barre. In smaller towns, a large proportion of the people subsisted on social security and welfare checks or money sent back home by family members working elsewhere. Towns in the southern fields became bedroom communities, their residents commuting daily more than one-hundred miles to steel mills in Bethlehem and Reading and government jobs in Harrisburg.

For miners too old to make a new start, there was enforced idleness: long days rocking on the front porches, a few dime beers in the corner saloon, walks along town streets lined with shuttered stores. Many could not come to terms with the waste of their lives. Stories abound of men who dressed every morning for descent into the mines, who, after an idle day at the fraternal hall, would return home claiming they had worked a full shift. Proud men who worked all their lives to support families watched wives and daughters go off to shirt factories; for such men dependency came as a degrading shock. Some cracked under the strain. A Pittston miner, concluding that he was useless, sat on a keg of powder and ignited it with his pipe. Many slipped into apathy, waiting for death.

The kingdom of coal is gone. The black rock that broke America's dependency on foreign coal, fueled an industrial revolution, kept millions warm, created great wealth, and birth to a vibrant immigrant culture has served its time in history. Anthracite's final legacy is a warning to all Americans that human lives and natural resources are finite and precious, that they can no longer be sacrificed indiscriminately on the alter of private greed. (48)

> Forty years I worked with pick and drill
> Down in the mines against my will;
> The coal kings slave, but now it passed,
> Thank be to God I am free at last.

This is from the tombstone of an anthracite miner in St. Gabriel's Cemetery, Hazleton, Pennsylvania.

Chapter Four

A SHORT CHILDHOOD

St. Nicholas breaker, near Mahanoy City, Pa. (l)

I spent my early years in St. Nicholas, a small coal-mining patch in Schuylkill County. It was somewhat different from any community one might reside in today. It had different smells, different noises, and different surroundings. Everything around me seemed so big as a young lad, maybe because everything was. Behind our company-owned clapboard house sat a culm bank that seemed to want to touch the sky. It was as if man and the creator were having a contest to see who could build the tallest mountain. That huge culm bank was a result of the coal waste being generated from processing coal in the breaker that lay directly behind it. The St. Nicholas Breaker was the largest coal breaker in the world. In front of our house, across the road, was the railroad yard. Resting in the yard were rows and rows of loaded coal cars, sometimes in the hundreds, waiting their turn to be hauled away. There was always a lot of activity in my neighborhood, large coal trucks bringing their loads of coal to the breaker for processing, massive steam engines moving rail cars to their proper locations. It was almost as if those monstrous engines were alive. They would grunt and groan, screech and chug, and even whistle. When they were getting a drink at the water tower, you would hear them *hiss.*

The sights and sounds of my early years growing up in the patch are certainly fond memories.

Today, the patch is gone, the breaker sits idle, except for the birds and bats that find it to be a good shelter it is useful no more. Time's change!

We can go back to the same area a generation before me and see how those young lads spent their days.

After about six-years of age their childhood was over. They went to work. The cycle of a miner's life began early. Every fourth worker was a boy. He usually started out in the breaker, sometimes as young as six, but normally at eight or nine. The breaker was a towering structure that loomed one-hundred to one-hundred-fifty feet above the other colliery buildings. It was the most important building and its peculiar design made it unlike

Breaker boys at their work, Eagle Hill Colliery (m)

any others. Its architectural form followed its function. Coal brought up from below was brought up to the top of the breaker in mine cars along on inclined plane. The coal was tipped from the cars into revolving cylinders that crushed and screened the coal, separating it into various sizes. The coal was then fed downward into a series of chutes. Boys sat crouched on narrow wooden planks over the moving coal, their feet in the chute to slow the flow of the coal. Below them were other boys each responsible for picking out the slate and refuse missed by those above. Sometimes old men and injured miners who could no longer work underground worked as "breaker boys". Hence, the popular refrain, "twice a boy and once a man is the poor miner's life."

In the industry's early days the breaker boy's workday averaged ten hours a day, a workweek was six days. The daily wage was forty-five cents. Before water was used in cleaning and processing the screening room, where slate was picked, was so thick with dust that the boys could hardly see beyond their reaches. They wore handkerchiefs over their noses and mouths and often chewed tobacco to keep from choking. Later mechanical breakers were somewhat cleaner, but the noise was horrendous and the whole building shook from the movement of the belts and chains that pushed the coal along. The windows in these breakers often were broken, and in winter icy air swept in, while in summer stifling heat caked coal dust on sweat-soaked bodies. Even after gloves were introduced, many companies forbade them because it impaired their sense of touch. This resulted in bleeding fingers, a condition known as "redtop". Old timers claim that the drops of blood in the snow could follow the path the boys took home after work.

Over the Coals, a poem by James Sweeny Boyle, catches something of the pathos of their lives:

> Over the ice, they pull the coals,
> Their fingers rent by a hundred holes
> You may trace the path torn digits tread
> By the crimson stream on the iced chutes shed.
>
> Their heads are bowed and their bodies cramped,
> A painful look on their features stamped
> Their knees are pressed against aching breasts
> Till bones are bent in their chambers tight.
>
> In sorrow they slave where the massive screen rolls,
> So fearfully, tearfully over the coals.

Breaker boys at the Eagle Hill Colliery, P&R Coal & Iron Co.
New Philadelphia Pa. (n)

In the deep snows of winter, fathers carried smaller boys to the breakers on their backs in the predawn darkness, or mothers would take the younger boys and return to wait for them at the end of their shift.

A story from the World War I era in Wilkes-Barre, relates how a mother carried her son's lunch pail to the breaker for him because he was too small to handle it himself.

After steam-powered mechanical chutes were introduced, many boys who fell into the chutes were killed, their bodies horribly mangled by the rollers. There is a ballad about a mother who went insane when her son was crushed by the rollers in the Audenreid breaker near Hazleton. She returned to the breaker day after day, always walking home alone. (49)

> Mickey Pick-Slate, early and late,
> That was this little breaker boys' fate
> A poor simple woman at the breaker still waits,
> To take home her Mickey Pick-Slate.

Manus McHugh was another boy who lost his life in the breaker. His job was to oil the breaker machinery. At noontime, one day, he was in a hurry because he wanted to get outside and play with the others. Rather than taking the time to shut, the machinery down he decided to oil it while it was still in motion. His arm was caught in the gears and he became hopelessly entangled.

After an investigation of McHugh's death, a 1903 report stated, "Boys will be boys and play unless they are held under by strict discipline", other than that, no changes were made to make conditions safer for the boys. (50)

At the end of their shifts, the boys clambered down the rickety stairs of the breakers, happy to be free, their teeth gleaming in coal-dust-blackened faces. After a wash and supper, they sometimes ran off to night school, where a single teacher taught all ages, with light coming from a lamp carried by each boy. In some schools, the history and geography lessons were sung in improvised verses that the teacher invented. Most teachers, male and female, seldom lasted long in one school. They usually were not tough enough to deal with the boys.

Work discipline in the breakers, was enforced by foremen who used clubs or leather switches to keep the boys at their work and enforce order. Whippings and similar harsh treatment frequently resulted in spontaneous strikes, slowdowns, or sit-downs. However, usually these strikes were short-lived as the bosses went after the boys with whips, a practice known as

Breaker boys picking slate in a coal breaker in 1892, near Pottsville, Pa. (o)

"whipping them in". Still, the boys remained contemptuous and rebellious, shouting and swearing like troopers above the noise as they developed the independence and scorn for the bosses that they would later show in the mines. (51)

Breaker boys were used in the minefields until well into the twentieth century. The Pennsylvania legislature, in 1885, made it illegal to employ boys under fourteen years old inside the mines or under twelve years old on the surface. In 1903 these limits were raised to sixteen and fourteen, respectively but the laws were seldom enforced. Parents eager for additional income filed false affidavits on a boy's age with local magistrates, who collected a twenty-five-cent fee for each document processed. The companies did not object because they paid the boys so little.

It was extremely difficult to determine the exact number of boys and their ages because seventy-five percent of them were foreign born. We do know, however, that as of 1905, seventy-five percent of the slate pickers killed were under sixteen years of age. (52)

After about five or six years working as a breaker boy, he would often go underground as a nipper. The nipper or door tender was the youngest of the underground boys, usually eleven to thirteen years old, his job was to open and close the heavy wooden doors that were constructed across the gangways of the mine.

The doors were an important part of the mines ventilation system. Above ground, huge fans forced fresh air into the mines. When the doors were closed, the air hit the doors and turned into the tunnels and chambers where the miners were working. The flow of air also pushed out dangerous gasses, like methane, so they didn't build up in the pockets and explode.

As the nipper sat on his bench outside the door, he listened for the rumbling approach of empty cars heading into the chambers or full cars coming out. Sometimes mules pulled the cars through the doorways, other times, in sloped areas, the cars rolled by gravity alone. When the nipper heard the cars, he opened the doors to let them pass through, and then made sure that the door closed tightly behind them.

It was a long day for the nippers who became bored sitting alone in the dark with only a little glow from his carbide lamp. The nipper knows that falling asleep could be disastrous. If the doors weren't opened, the racing mine cars, which weighed about four tons when fully loaded, would crash into the doors. In 1903, a young nipper who fell asleep behind the gangway jumped up when he heard the approaching mine cars. He went to throw open the doors but it was too late, the mine car struck and killed him.

The darkness and solitude bred courage and responsibility in the boys. Creaks, groans and trickling sounds forewarned that a roof was "working", which meant that the rock was loose overhead and it was in danger of collapse. If a roof was working in a gangway or heading, the nipper was usually the first to hear the telltale sound and could run to warn the others.

Mule drivers, nippers (door boys) and spraggers at the close of a day in a mine, 1915 (p)

Another job a boy might have in the mines, if he had the skills, was a spragger. Only the fastest and most agile boys were picked to be spraggers. Their job was to control the speed of the mine cars as they rolled down the slope. The spragger ran along the side of the cars and jammed sprangs (wooden sticks) into the wheels of each car. The sprags acted as brakes, locking the wheels in place and slowing the cars down.

Spragging was dangerous work. The boys dodged low ceilings and passageways that narrowed without warning to clearances of only several inches. At times, spraggers hands and fingers got too close to the wheels and became wedged and sheared off.

If the wheels weren't spruge properly, the cars flew out of control. Often they toppled or jumped the track and crashed into the mine wall sometimes injuring or killing the spragger.

The boys loved the danger and excitement of their work. At least they weren't hunched over long chutes of coal or perched on benches outside the gangway doors. Here, in the miles of tunnels, they even found some opportunities to have a little fun. "We'd stand on the bumper and ride the cars down the slope," said Richard Owens. "It was pretty lively going down that way, of course, the bosses didn't like it."

The bosses didn't like it because nearly forty-percent of fatal accidents happened when the boys riding the cars were struck by a low ceiling or when they fell off and were run over or dragged by the cars.

Young underground mineworker (q)

The most glorious job of all was that of a mule driver. It offered danger, excitement, and best of all, freedom to move about the mines.

The mule driver, usually a boy in his early teens, traveled from one work chamber to the next, coupling the full cars and leaving an empty car to be filled. A boy started out with one mule, and then worked his way up to a six-mule team. When he was able to drive six mules, he was given a man's wage and earned the respect of all the workers and bosses.

Because of the narrow passageways, the mules were harnessed in tandem, one behind the other. The driver stood on the front car bumper, where he used only his commands to guild the mules. If the mules were stubborn and refused to move, he cracked a warning in the air with his long, black, braided, snake whip. Most foremen didn't allow the boys to carry watches because they wanted the drivers to concentrate on the number of cars, not the time. If there was a lull in work, the driver practiced his skill with the whip. He would set up the mine lamp and with the crack of his whip tried to extinguish the flame without upsetting the lamp.

Mule driver (r)

During the West Pittston Mine fire in 1871, Martin Crahan, a twelve-year-old mule driver, refused to ride the cage to safety because he knew that there were nineteen miners that needed to be warned of the fire.

Martin found the miners, told them of the fire, and then ran back to the elevator cage. It was too late; the cage had already been destroyed. Realizing there was no hope for rescue, he found his way back to the stable. He found his mule and scratched a final message to his family then he died next to his mule.

On Easter Monday, in 1879, mule driver William McKinney told his mother not to pack much lunch for him because he planned to be home early. He snuck extra lumps of sugar into his pocket for his mule, Harry.

At work, as William and Harry plugged along the gangway, William saw rats heading up the slope. A few seconds later, he heard a terrible crash and a gust of wind and dust pushed the mule back and knocked William off the mine car.

A large cave in had crushed the gangway and the airway. "I thought of the men inside," said William. "I knew the air was shut off and that some would be burned or blown up if we did not go down to warn them. So I gave Harry a jab in the ribs and away we went."

William found the men and warned them and at that point another cave in occurred. "It was like a big thunderstorm with timbers cracking and splitting until broken. We were trapped. Everything around us was sealed tight. Harry and me were alone with seven other men."

For three days, William and the men waited for rescue. As they sat, they talked about their families and friends. Their lunches were gone and all they had was two gallons of oil and four safety lamps. By the fourth day, with no sign of help, and with nothing to eat, their eyes seemed to be bulging out of their sockets, and they looked wild. Finally, one of the men looked at Harry and said, "I think if I had a piece of that mule, I'd eat it." After much thought, William said, "I'll bring the mule here but I won't kill it, one of you will have to."

As William held Harry by the bridle, Harry sniffed William's pockets for sugar. Thru tears, William told his mule, "Harry, I guess I won't be giving you any more sugar."

Harry was killed, cut up and fried on the lid of a lunch pail. After nine days, rescuers found the entombed men. When William came out his mother was so overjoyed to see him that she gave him the new suit that she planed to bury him in. (53)

So many of these young lads went from the cradle to the grave with not much in between and they accepted each new challenge in life without complaint. Going from breaker boy, to nipper, to spragger, to mule driver and finally to miner was as normal for these young lads as kindergarten, grade school, middle school, high school, college and then gainful employment is for a youngster today.

The odds that some of these boys would reach their senior years were not very good. But even if they did, there were no pensions, no healthcare, no social security, and no nest egg. Their later years would be as dismal as the years before.

Here's a lyrical verse telling of an elderly miner whose next journey in life would be taken with the angels. He reminisces about his days in the mines and remembers the boys whose journey has ended, the boys that he worked with in his youth.

I'm getting old and feeble
And cannot work no more.
I have laid my rusty
Mining tools away.

For forty years and over
I have toiled about the mines.
But now I'm getting feeble
Old and gray.

I started in the breaker
And went back to it again.
But now my work is finished
For all time.

The only place that's left
Me is the alms house
For a home.
That's where I'll lay this weary head of mine.

Where are the boys who
Worked with me in the breaker
Long ago. There are many of them
Now have gone to rest.

Their cares of life are over.
They left this world abode,
And their spirits are now
Roaming with the blessed.

In the chutes I graduated,
Instead of going to school;
Remember friends my
Parents they were poor.

When a boy left the cradle,
It was always made the rule,
To try to keep starvation
From the door.

At eight years of age,
To the breaker, first I went
To learn the occupation of a slave.

I was certainly delighted,
And on picking slate was bent;
My ambition it was noble,
Strong and brave

I next became a driver,
And though myself a man.
The boss he raised
My pay as I advanced.

In going through the gangways
With the mules at my command,
I was prouder than the President of France.

But now my pride is weakened
And I am weakened too.
I tremble 'till I'm scarcely fit to stand.

If I were taught book learning,
Instead of driving teams;
Today my friends
I'd be a richer man.

I next became a miner,
And laborer combined.
For to earn my daily bread
Beneath the ground.

I preformed the acts of labor,
Which came in a miner's line
For to get my cars and load them
I was bound.

But now I can't work no more.
My cares of life are done.
I am waiting for the signal at the door.

When the angels they will whisper,
Dear old miner you must come,
And we'll roll you to the bright celestial door. (54)

Chapter Five

SOLIDARITY COMES TO THE COAL FIELDS

One of our great Presidents' Abraham Lincoln once stated that, "Our fathers brought fourth on this continent a new nation conceived in liberty and dedicated to the proposition that all men all created equal." True, all men are created equal, but in this in-perfect world we live in, sometimes all men aren't treated equal. In the anthracite coal region of northeastern Pennsylvania, this was a common practice, coal barrens reaping huge profits at the expense of the destitute immigrant mineworker.

The one unifying experience in the lives of the miners was their work. Mining was the most dangerous job of the day. In the residential blocks of many mining communities, many families had lost some member in the mines: father, son or brother. To the hazards underground were added exploitations of other kinds. Miners were paid for the coal they dug, by the cubic-foot, carload, or ton. Possibilities for cheating abounded. Unscrupulous company employees "short-weighed" miners production or claimed there was too much slate mixed in with the coal. The miners paid for their own helpers, powder, tools and the supplies they used. The operators forced the miners to buy materials from company stores as a condition of employment, at prices fixed by the operators. When deductions were made for helpers' pay, supplies, and rent for the company owned house a miner could take home less than one-forth of what he had earned for the amount of coal actually dug. (68)

I have heard the old-timers say that the mule in the mine was worth more than the miner. If the miner was killed or injured, you could always hire another one, but if the same fate befell the mule, it would cut into the company's profits because a new one would have to be purchased.

The only way a miner could fight back was to organize. Alone the mineworker had no power over the coal companies, but coming together as one in the form of a union, they could stand up and be heard.

Early attempts at unionization in the coalfields, however, resulted in multiple failures. In 1835, boatmen on the Schuylkill Canal formed a

Committee of Vigilance and struck, but this organization soon disbanded. In 1842, the first recorded strike in the anthracite fields broke out and involved more than two-thousand men, but it was a spontaneous effort that failed to produce a permanent union. Then in 1849, John Bates organized what became known as the Bates Union and called out two-thousand men to protest the payment of script, redeemable only at company stores. The Bates Union lasted less than a year, to be replaced by ineffective local unions and associations. (69)

Even though some of the early unions failed, they were small victories for the mineworkers. It made the coal companies aware of the mineworker's willingness to stand up and be heard. A work stoppage, even for a day, would result in huge financial losses for the coal companies.

The breaker boys did not have much schooling, but they learned their lessons in life well. The young lads that worked at the Muskrat Breaker near Moosic were willing to shut down the whole operation in order to have one of their grievances recognized.

The barbarous methods backfired on a breaker boss with a wooden leg, the boss, whose name was Bill, had forty or fifty boys under him. As an old-timer who worked for him said, "And when I say that he had boys under him, I mean just that! For a man with a wooden leg, he could skip and hop over those seats and chutes with the speed and accuracy of a squirrel flying from tree branch to tree branch. And if he found a single piece of rock or slate in any chute about to enter the coal pocket, he proceeded in a methodical and efficient manner to make every boy on that chute of coal realize the grave necessity of clean coal." Bill's methods were cruel and unique. "He would raise that wooden stump and give each boy a prod in the back, or use it as a club to inflict punishment on little backs already aching from constant bending of the body above the chutes of coal, or he would come up from behind a boy and take him by both ears and lift him a foot or two above the seat."

One sunny July afternoon, after work, the boys met at a near-by swimming hole and decided that they would spend the following day swimming rather than report for work. The next day all the boys, except for three or four, who were promptly denounced as scabs, assembled at the swimming hole. The furious Bill, the colliery superintendent, and the outside foreman, soon confronted them. The bosses stood on a natural rock overhang above the swimming-hole and demanded that the boys return to work. While the bosses were exhorting the boys, two of them suddenly rushed up on them from behind a clump of bushes. They ran into the unsuspecting Bill and went sailing off the ledge with him into the water. As the three bodies hit the water, with a mighty splash, the other boys dove into the water and surrounded the puffing, sputtering breaker boss. They dove under him and pulled him down, and when he reached the surface, gasping for air, they splashed water in his face. It took a full fifteen minutes before the other two bosses could pull the half-drowned Bill from the swimming hole.

Negotiation then began. The boys stationed themselves on the other side of the swimming-hole, facing the bosses, and demanded that as a condition for their return to work Bill is relieved from his duties as breaker boss. After a while, the superintendent finally gave in to their demands and the boys triumphantly returned to their places on the coal chutes. (70)

In the twenty years following the Bates union, the coal region was plagued with murder, assaults and arson, mostly blamed on the Irish Catholic mineworkers of the region. This was during the Molly Maguire era.

This time the Irish had nowhere to run and no desire to do so. Instead, they stood proud and ready to issue a call for change in the mining industry. This battle cry brought forth a new list of Irish heroes who led the charge to change the industry forever. One of the earliest of these Irish warriors in America was John Siney. In 1868, he formed the first successful anthracite labor union and christened it the Workingmen's Benevolent Association (WBA). For a few years, the WBA was an active force in improving conditions.

The WBA quickly organized a committee on political action to direct its membership in a battle for improved conditions in the mines. This committee traveled to Harrisburg to lobby successfully for the first mine safety law. The act as passed was designed to provide, "for the better regulation and ventilation of mines and for the protection of the miners in the County of Schuylkill." The act of 1869, held many requirements that would be the basis of the mine safety legislation for the future. The fact became painfully evident on September 6, 1869, when one-hundred-ten men and boys died in the Avondale mine disaster. Siney responded to the disaster by working to bring miners together for change. He made this statement, "You can do nothing to win these dead back to life, but you can help me to win fair treatment and justice for the living men who risk life and health in their daily toil." Under the urge of Siney, the general Council of the WBA sent a committee made up of members from each county union to Harrisburg to demand better mining legislation. Stunned by the public outcry that followed the Avondale disaster, the Pennsylvania lawmakers quickly passed a more detailed Mine Safety Act of 1870

Although the gains it had supported were impressive, the WBA would suffer greatly over the next half-decade. An extended strike in 1875 forcing miners back to work on operator' terms and the hanging of twenty Irish mine workers (Molly Maguires) brought an end to the Workingmen's Benevolent Association that was established by John Siney.

An illustration from Leslie's Popular Monthly for March 1871, showing scabs, or blacklegs, being taunted by a crowd of miners and their wives. Mahanoy City, Pennsylvania, 1871. Courtesy of the Library of Congress.

(ee)

Yet the seeds he had sown with the support of the earliest mining legislation would grow to become a garden of change as resilient as a field of Irish clover. (71)

From 1875 until the end of the 1890s, organized labor had virtually no representation in the anthracite fields. Efforts were made by the Knights of Labor to organize workers in the late 1870s to 1888. The Knights organized sixty local assemblies in 1877, but did not have much success in negotiating with the operators.

In 1894, the United Mine Workers of America (UMWA) organized their first locals in the Schuylkill fields. Membership in the United Mine Workers was small in the beginning.

As membership grew, so did opposition from the coal companies. Unfortunately, violence was not foreign to the coalfields. One such incident of violence and death was the Massacre at Lattimer Mines on September 10, 1897.

The small coal-mining patch called Lattimer Mines is located near Hazleton, and was populated mostly by Italians. The surrounding patch towns consisted mainly of Slavic immigrants. They lived in company homes built and rented out by mine owner Ariovistus Padee of Hazleton, one of the wealthiest individuals in the United States at that time.

While conflict between mine owners and mineworkers was common in the coal region, the summer and fall of 1897 proved to be a particularly troublesome period.

Immigrant workers had been agitated for sometime for a variety of reasons. For one, coal companies often adopted paternalistic attitudes toward immigrants. Some operators believed that the company's wealth and power gave them the right to use and treat workers as they wished. Many immigrants experienced prejudice and bigotry.

Through the summer of 1897 policies targeted against immigrant workers continued. On Monday, August 16, more than three-hundred-fifty angry workers, protesting the violent attack on a young picketer at the Honey Brook Colliery in McAdoo, marched to each of the neighboring Lehigh and Wilkes-Barre Collieries closing them down. By the end of the day, more than three thousand workers joined the strike. By now, Mining communities were in the state of turmoil. Worker anger was building, strike activities expanded, UMWA organizing continued as more and more workers voted to affiliate with the union and minor outbreaks of violence became more common.

On September 1, Lehigh and Wilkes-Barre workers formally voted to strike. The walkout soon swelled to five thousand and included employees of smaller independent coal companies. Marches were organized with the intent to shut down collieries. On September 3, a group of more than a thousand men proceeded from McAdoo to Hazleton and closed down several collieries. Demonstrations continued throughout Labor Day weekend and by mid-week some ten-thousand workers from throughout lower Luzerne County and neighboring Carbon and Schuylkill Counties joined in the protest.

Coal company operators thought they had a war on their hands. In their view, immigrant miners were acting like vigilante's bent on attaining their goals. The operators wanted to put an end to what they considered lawlessness. Luzerne County Sheriff, James Martin, vacationing in Atlantic City over the holiday, was called back home. His expertise was demanded by the coal operators to quell the growing unrest. It was up to Martin to end the strike.

Martin declared a state of civil disorder, which authorized him to form a posse. By the evening of September 6, he had over eighty volunteers. Martin swore in his deputies and armed them with new Winchester rifles, accompanied by metal-piercing bullets.

On the evening of Thursday, September 9, a delegation of Lattimer workers met with striking UMWA members in nearby Harwood. The Lattimer miners wanted to join the walkout and requested that the strikers march to Lattimer Mines the following day to close down the colliery. Knowing that no concessions could be won from Pardee without a show of unity, the strikers agreed to aid their Lattimer associates and the march was planned for the next day.

Friday, September 10, was a warm summer day. Some three hundred men assembled at Harwood. They gathered a few American flags to display during their march and set off for Lattimer. They proceeded peacefully and unarmed. In their view, they were simply expressing the American rights to assembly and free speech.

Sheriff Martin received word of the procession and mobilized his deputies. As the column neared Hazleton, they encountered Martin's posse. The sheriff drew his pistol, pointed it at the head of a marcher, and ordered them to disperse. They refused. A fight broke out. One of the deputies grabbed a flag and ripped it to pieces. Further violence was averted after the police chief insisted that the column could continue only if they agreed to march around Hazleton. They agreed and proceeded peaceably. Anxieties were running high.

Strikers on their way to Lattimer Mines the day of the massacre.
September 10, 1897 (ff)

Martin and his deputies boarded trolleys to pursue the marchers at Lattimer. Trolley passengers reported that the talk was of a shooting. One deputy was overheard saying; "I bet I'll drop six of them when I get over there" A reporter relayed to the *Wilkes-Barre Times* that serious trouble was on the horizon. Word spread quickly. In Lattimer, children were hustled from the schoolhouse by apprehensive mothers. The colliery whistle sounded a shutdown. Company police and other deputies met Martin at Lattimer with a combined force of over one hundred fifty men. They lined the forked entrance to the town and waited.

At nearly three forty-five in the afternoon, the marchers, now numbering over four hundred, approached Lattimer with the American flag in the lead. Martin walked to the head of the column and announced that they must disperse. Not all marchers, particularly those in the back, could hear him. Martin attempted to tear the flag from the hands of Steve Jurich. Thwarted, he then grabbed a marcher from the second row. When others came to his aid, a scuffle broke out while part of the group continued forward. Martin drew his pistol and pulled the trigger, but the weapon did not fire. Then someone yelled, "Fire, give two or three shots!" Several witnesses claimed it was the sheriff, though he would later deny this. A barrage of shots rang out. The flag bearer was the first man hit. He cried out to God in Slovak as he fell, mortally wounded. Several marchers at the front of the column realized that the deputies were not using blanks. Those who understood what was occurring immediately began to scatter. Some ran toward the nearby schoolhouse. Teachers Charles Guscott and Grace Coyle watched the events unfold and thought the first bullets were blanks, until they saw several men running toward them fall to the ground. Other shots pierced the schoolhouse walls, sending wooded splinters flying through the air. Some deputies broke rank to take better aim at the fleeing marchers, shooting them in the back as they ran trying to escape the bullets. Miner John Terri threw himself on the ground. Another miner fell on top of him, dead. Andrew Jerecheck attempted to run toward the schoolhouse and was stopped by a bullet in the back. He pleaded in vain that he wanted to see his wife before he died. Mathias Czaja was likewise hit in the back and fell to the ground. Some of the wounded cried out for help to which one eyewitness heard a deputy respond, "We'll give you hell, not water, you hunkies."

The shooting continued for at least a minute and a half, though some eyewitnesses claimed it might have been three minutes or more. Perhaps as many as one hundred-fifty shots were fired. The magazines in several of the sixteen-cartridge Winchesters were fully discharged. Blood, smoke, road dust, and cries of anguish overwhelmed the scene. Nineteen marchers lay dead. Another thirty-six were wounded. The force of the steel piercing bullets literally tore many of the bodies to pieces. Even those who had taken bullets in their limbs were critically wounded. A few of the deputies walked among the dead and dying kicking them, while others helped those who were wounded. When the shooting stopped Sheriff Martin uttered, "I'm not well."

News of the bloodshed spread quickly, wagons and trolleys moved the dead and dying to local hospitals and morgues. While Sheriff Martin departed for Wilkes-Barre to meet with his attorney. Families of the marchers gathered in anguish and disbelief to learn the fate of the men. The deputies scattered to Atlantic City to seek refuge under assumed names in the Traymore Hotel.

By the next day, the Governor detached the Third Brigade of the State Militia to the Hazleton area to maintain public order, as it was feared that reprisals for the killings were all but certain. However, except for one attack on the home of a mine superintendent, the immigrants remained peaceful in their grief, hoping that the American court system might bring the deputies to justice. Funerals continued for several days, sometimes drawing crowds as many as eight thousand. Polish, Slovak, Lithuanian and

Pennsylvania National Guard, Third Brigade, in Hazleton
1897 (gg)

other ethnic organizations, regionally and nationally, expressed their grief and outrage at the Massacre at Lattimer Mines.

In late February 1898, Sheriff Martin and his deputies were tried for murder at the Luzerne County Court House in Wilkes-Barre. The trial, which lasted twenty-seven days, ended in an acquittal. While the posse walked free, resentment and bitterness were not as easy to snuff out, as were the lives at Lattimer Mines.

Deep scars remained from the events at Lattimer, particularly among ethnic groups in the anthracite region. While the outcome was tragic, in the longer term, the 1897 strike and the Lattimer Massacre secured the role of immigrant workers as a significant element in America's rapidly unfolding industrial order. It was clear that those who adopted America as their home were not willing to remain docile in the face of nearly intolerable living and working conditions. In continued efforts to improve their lives in a land which held great promise, those who mined anthracite did not abandon collective action to address their grievances. They joined the UMWA in great numbers and it became the single most powerful representative of the anthracite workers as its numbers in the coal region swelled to well over one hundred thousand. (72)

Twenty-nine year-old John Mitchell was elected President of the UMWA in 1899. An Irish immigrant, young John Mitchell knew of the miners concerns because he walked in their footsteps. John entered the mines when he was twelve years old. He rose quickly in the mines, from door boy to miner's helper. Though slight of body, he soon developed the miner's toughness. And no one doubted his courage. After a mine disaster that he helped in the rescue operations, another miner remembered seeing a trail of blood on the ice left by Mitchell's cut bare feet. (73)

The miners in the anthracite coal region took an immediate liking to Mitchell. He was especially popular among the immigrants, most of all because he listened to them and seemed directly interested in their welfare. In 1901 when news spread throughout the anthracite region that the president had been shot, crowds of immigrants gathered crying, "Who shot our President!" However, they were dispersed when they learned that President Mitchell had not been shot but only President McKinley. (74)

The anthracite coal miners were always willing to stand up to the coal operators, but their one stumbling block, in their quest for unity, was their division caused by their ethnic backgrounds. With one statement, John Mitchell removed this obstacle. At a union rally Mitchell stated, "The coal you dig isn't Slavish or Polish or Irish coal, it's coal." This one phrase transformed the immigrant coal miners into American coal miners and with that, they could come together as one.

In the summer of 1902, the anthracite coal miners went on strike for over one hundred-sixty days and sent the nation into a panic over a possible coal shortage, prompting Presidential intervention.

John Mitchell "boy president" of the UMWA (hh)

In order to understand the strike of 1902, one must consider the events in the years proceeding. The main contributor to the strike of 1902 was the anthracite strike of 1900. John Mitchell attempted to bargain with the coal operators for recognition of the union, improvements in wages and hours, and better working conditions. The main bargaining point for the coal miners was for an increase in wages, because wages in the anthracite fields had been stagnant.

The coal operators, however, refused to negotiate with Mitchell. They also refused to recognize that the union existed. Because of the coal operator's refusal to bargain, John Mitchell called a strike of the anthracite miners in September of 1900. Unfortunately, for the coal miners, 1900 was an election year, and President William McKinley was running for re-election. Representatives of McKinley's party were afraid that the strike might interfere with his election and ordered Republican representatives to end the strike. Because of the political pressure, the strike was negotiated and ended without most of the miner's demands being met. In the settlement, the miners were granted a ten-percent raise, but the UMWA was not recognized as the bargaining agent and representative of the anthracite coal miners. This unresolved issue would eventually help to lead to the strike of 1902.

The ten-percent wage increase won from the strike of 1900 was only temporary in appeasing the demands of the miners. Working conditions and hours were not improved and the union was still not recognized as an official representative. It was only a matter of time before the anthracite miners went on strike again. On May 12, 1902, the anthracite coal miners walked off the coalfields, demanding more wages, an eight-hour work day, and recognition of the United Mine Workers Union. The UMWA representatives attempted, once again, to negotiate with the coal operators, but once again, they refused. The strike continued for over five months before further outside action was taken.

As the strike continued into October, and the winter months rapidly approached, citizens were becoming very concerned about the possible coal shortage during the winter. President Theodore Roosevelt was also becoming concerned and decided to take unprecedented action. President Roosevelt invited representatives of the UMWA and the coal operators to the White House on October 3, 1902, becoming the first President to personally intervene in a labor dispute. President Roosevelt reiterated the concerns of the American public that was being affected by the shortage of coal. The UMWA President, John Mitchell, agreed to call off the strike if a tribunal of presidential representatives, UMWA representatives, and coal company operators could be assigned to continue to deal with the issues of the strike, such as union recognition. Mitchell also asked for a small increase to the miner's wages until the tribunal had time to work out an agreement. The public saw the efforts of Mitchell to be noble and fair. The coal operators did not see this agreement as fair and once again refused to deal with the union, despite the pleas of President Roosevelt

After the meeting with the President and the coal operators, John Mitchell met with the coal miners. He discussed with them the concerns of the American public that President Roosevelt had raised during their discussions. Mitchell debated with the coal miners whether they should temporarily return to work in order to prevent a coal shortage during the cold winter months. The coal miners, however, voted almost unanimously to continue the strike no matter what the cost. They did not want to bow to political pressure, like during the strike of 1900, and suffer another defeat. Seeing that neither side was willing to back down, President Roosevelt had to take serious action again. He threatened to send military forces to take over and operate the anthracite mines. If this happened, coal companies would lose money, as well as the miners. Both sides were now willing to try to reach a compromise.

President Roosevelt appointed a commission to arbitrate the negotiations. Representatives from both sides met with the commission and agreed to follow their recommendations for ending the strike. On October 23, 1902, the coal miners went back to work, and the nation breathed a deep sigh of relief. The coal miners achieved a ten-percent wage increase and a reduction in the hours of the workday. Once again, however, the union was not recognized in the agreement as a bargaining agent for the coal miners. (75)

The great Anthracite Coal Strike of 1902 was important in setting a new precedent of presidential intervention. President Roosevelt personally tried to intervene and end the strike on behalf of the American public. Although his attempt initially failed, he was seen in the public's eyes as an understanding and concerned President. The public also saw the coal operators in a negative fashion because they were seen as the ones that would not compromise with the President. This helped to increase public support of the coal miners and the United Mine Workers Union. This soon became the trend of public opinion towards unions. (76)

John Mitchell during the Great Strike of 1902, being flanked by the breaker boys. (ii)

To the mineworkers Mitchell became an authentic hero; they voted October 29 a holiday and named it *Mitchell Day*. The anthracite workers remembered him only as their beloved "Johnny," "John d'Mitch," "Father." A delegation of Slavic workers had once told him "Blessed be the day . . . when you came amongst us." For years, after he left the coal region, Mitchell's photograph hung in many miners' homes along side a picture of the Sacred Heart. These people would remember him as the only president they really knew or cared about. (77)

The breaker boys were especially touching in their appreciation. When miners had attempted to keep the breaker boys away from union meetings, Mitchell had insisted upon their right to attend because they worked hard and long hours and were exploited even more shamelessly then the men. Mitchell had said of them, "They have the bodies and faces of boys but they came to meetings where I spoke and stood as still as men and listened to every word. I was shocked and amazed as I saw those eager eyes peering at me from eager little faces, the fight had a new meaning for me, and I felt that I was fighting for the boys, fighting a battle for innocent childhood." Before Mitchell left the region, twenty thousand breaker boys paraded before him in an unprecedented show of gratitude. They presented him with a gold medallion. He responded by telling them, "that their terrible lot had made his work in the anthracite fields a crusade." (78)

Despite the threat of physical harm and economic ruin, miners have constantly struggled against great odds to achieve their goals, the eight hour work day in 1902, collective bargaining rights in 1933, health and retirement benefits in 1946, and health and safety protection in 1969.

The UMWA was an influential member of the American Federation of Labor (AFL) and was a driving force behind the creation of the Congress of Industrial Organizations (CIO). Organizers from the UMWA fanned out across the country in 1933, to organize all coal miners, after the passage of the National Industrial Recovery Act. The law granted workers the right to form unions and bargain collectively with their employers.

After organizing the nations coal fields, the miners turned their attention to the mass production industries, such as steel and automobiles, and helped those workers organize. Nearly four million new workers were organized in less than two years.

The UMWA was an early pioneer of health and retirement benefits. In 1946, in a contract between the UMWA and the Federal Government, a multi-employer UMWA welfare and retirement fund would permanently change health care in the coalfields of the nation. The UMWA fund built eight hospitals in Appalachia, established numerous clinics, and recruited young doctors to practice in the rural coalfield areas.

Because of the dust created in the mines, the UMWA was forced to become an expert in occupational lung diseases, such as silicosis and pneumoconiosis. In 1969, the UMWA convinced Congress to enact the landmark Federal Coal Mine Health and Safety Act. The law changed a number of mining practices to protect miners' safety and provide compensation for miners suffering from Black Lung Disease.

Today, the United Mine Workers of America continues its primary role of speaking out on behalf of the American coal miners (79)

As Americans, we have become accustomed to a lifestyle that is unparalleled anywhere in the world. Sometimes we take it for granted. One must remember, that today, we are the beneficiaries of the sacrifice of those miners who were willing to put their lives on the line in order that their children and their children's-children and our children could have a better existence.

Yes . . . All men are created equal. However, sometimes corporate greed creates deaf ears and workers must come together so their voices can be heard.

Chapter Six

THE MOLLY MAGUIRE ERA

During the 1860s and 1870s, the anthracite coal region of Pennsylvania was the site of much labor unrest, which resulted in violence in the coalfields. Most of the labor unrest and violence was blamed on the Irish Catholic mineworkers of the region. Those were dark days in the history of the hard coal region. The coal companies were out to squash the organizing of the mineworkers at all costs. One of their tactics resulted in the hanging of twenty Irish Catholic immigrants. These martyrs of labor were all Catholic, pro-union mine workers and members of the Ancient Order of Hibernians.

The Irish were not strangers to persecution, that's one of the reasons they left their homeland. The reign of oppression was nothing new. Since the middle of the twelfth century, the Irish had suffered under the British. In the sixteenth century, the English extended their domination, when in 1649 Oliver Cromwell swept Catholic Ireland with "fire and sword." In the decade of his invasion more than one-third of the population of Ireland died, most of them from starvation and disease this totaled almost three-quarters of a million people, somewhat less than the number that died in the Great Famine two centuries later. The English decimated the Irish landholding classes, and reduced most of the people to a condition of complete destitution.

In addition to the land confiscation, all priests were required to register their names and parishes under penalty of being branded with a hot iron. Public crosses, which stood in every village in Ireland, were torn down. All bishops were banished under penalty of being hanged, or drawn and quartered. Friars and monks were also banished. No Catholic chapel could have a belfry, a tower, or a steeple. Catholic pilgrimages were outlawed. Catholic schools were outlawed. Catholics were forbidden from marrying Protestants, and any priest caught performing such a marriage was sentenced to death. They could not bear arms, or own a horse worth more than five pounds and the Gaelic language was outlawed. Lord

Chancellor Bowes summed up the situation when he said, "The law does not suppose any such person to exist as an Irish Catholic." The Earl of Clare, his successor, called Catholics "the scum of the earth." (55)

By the 1800s, the Irish Catholics lead impoverished existences, many of them tenants of Anglo-landlords, some on land that was in their families' centuries before, yet still they were faithful to the church St. Patrick had brought them.

A land of opportunity was opening up in the New World. In America, a large deposit of natural resources was discovered and a huge workforce was in demand to mine this commodity called anthracite coal.

The Irish Catholics, reluctant to leave the land of their fathers, began to exodus to America, a land promising to yield new hope to them and their families. The unskilled, after generations living under a regime that denied them upward mobility, soon found work digging the canals, building the railroads and mining the coal of America.

Jobs in the mines were plentiful. Towns full of Irish immigrants sprung up in the hard coal region of Pennsylvania. At first, these jobs seemed like a blessing to the destitute immigrants. The coal companies provided company owned housing with rent deducted from the miner's paychecks and company owned stores, where a bill could be accumulated. However, the Irish soon discovered the trap. After the house rent and excessive prices

Miners in the Schuylkill field, the center of Molly Maguire activity (v)

at the company store were deducted from their paychecks, there was very little money left. They were trapped in a system that promised them only generations of survival at the poverty level. Their young children would have to work at the breakers for the families to survive. And woe to the family whose father was killed or crippled in the mines. Without an income to pay for the housing, the family was turned out to the muddy streets to beg or rely on friend's charity, much the same situation they had fled from during the Great Hunger. Any Irish miner speaking out against the conditions faced possible blacklisting among the growing cartel of mine owners, leaving him unable to find work in the region, almost guarantying starvation for him and his family. Justice in the coal region was administered by private police forces owned by mine operators.

In the days, long before the social programs of Franklin Roosevelt, the Ancient Order of Hibernians (AOH) administered its own welfare programs from dues and fundraisers. The AOH offered death benefits for widows and children, sick pay for those miners unable to work and funds for periods of unemployment. Nevertheless, little was available and there was so much need.

Perhaps taking a cue from the mine owners, who were consolidating their operations and presenting a solid front against their workers, the miners began to organize. What one man or even all the workers at one mining operation could not accomplish, perhaps an organization of workers across all the mining operations could. The action would effect not only the mining operation, but also the railroads that owned coal lands or relied on the coal trade for most of their freight. Men organizing into one body acting in concert across the coal region could apply pressure for better working conditions and better wages.

In 1868, John Siney, of St. Clair and a native of Ireland, formed the Workers Benevolent Association (WBA). The goal of the WBA meshed with those of the AOH, which was to take care of their fellow men, helping to improve their lives.

The handwriting was on the wall for the mine owners. Any improvement in the lives of the Irish would cut into their profits. The coal and railroad owners found an ally in Benjamin Bannan, editor of Pottsville's *Miners' Journal.* Bannan wrote editorials against Irish Catholics, their union and their political aspirations. It was Bannan who introduced the name Molly Maguire to America. Molly Maguire was a legendary figure in Ireland, an old woman who led secret attacks against absentee landlords' rent collectors. Through Bannan's editorials, the name Molly Maguire came to symbolize lawlessness and violence. Every crime in the coal region was blamed on the Molly Maguire's. By linking the violence in the coal region to a phantom group named Molly Maguire and then branding the AOH and the WBA as Molly Maguire's, Bannan and the coal operators hoped to break the power of the unions. (56)

Philadelphia & Reading Coal & Iron Police (w)

If the limited victories of the WBA raised some optimism among the miners, they generated considerable alarm among the operators. Though the smaller mine owners in the southern field might recognize the union and make concessions, the big companies were intransigently unwilling to give way. They lacked a plan of action and someone ruthless enough to carry it out, unfortunately for the miners, there was a man willing to do the job.

Franklin Benjamin Gowen became president of the Philadelphia and Reading Company, in 1870, at the age of thirty-three. During one of the mine strikes he tried to crush, he is reputed to have said, "I'll turn Schuylkill County into a howling wilderness before I give in to the miners," Between 1871 and 1875, Gowen and the Reading purchased over one-hundred-thousand acres of land, in the coal region, at the cost of forty-million dollars.

Control over the coal industry also required control over labor. Gowen knew that he had to crush the miners union and he set out to do so. He hired Allen Pinkerton, head of one of the first detective agencies in America. He gave him one-hundred-thousand dollars to infiltrate agents into the AOH and the alleged secret society known as the Molly Maguire's. (57)

Franklin B. Gowen, President of the P&R. Railroad (y)

One of the most noteworthy Pinkerton agents' was James McParlan, alias James McKenna.

Disguised as a tramp, McParlan left Philadelphia for the coal region on October 27, 1873. He quickly made acquaintances by his free-spending ways and learned something of the Mollies. He found out that their main strength apparently was centered in Pottsville, Shenandoah, Mahanoy City, and the surrounding "patches".

A chance remark sent the detective on to Pottsville. There, on a fine fall day, he lurched in feigned drunkenness through the doors of the Sheridan House, a saloon of considerable reputation owned by a gigantic Irishman—and reputed Molly leader—Patrick Dormer. While the regulars looked on suspiciously, McParlan broke into a sprightly jig to the accompaniment of a fiddler who was playing away in a corner. The skillful heel-and-toe routine did the trick. He was quickly set up with a shot of whiskey. He then touched the hearts of the good fellows with a lilting ballad from County Donegal—the ballad of Molly Maguire.

McParlan was then invited by Dormer to be his partner in a game of cards in the back room. The pair began to lose steadily, but the reason soon became obvious. McParlan grabbed the hand of one of his opponents, catching him using six cards instead of five. The cheater challenged the detective to a fistfight. The bar room was cleared; in five rounds McParlan soundly beat him. He then stood everybody to a round of drinks. The charmed Dormer sat the stranger down at a side table to find out more

71

about him. He said his name was James McKenna, and he was not a tramp at all but a fugitive on the run for a killing in Buffalo, New York. His money came in part from a pension fraudulently obtained for service in the U.S. Navy during the Civil War but mostly from counterfeiting or "passing the cheat." He also implied that he had been a member of the AOH in Ireland and New York State but suggested that any inquires might expose his whereabouts to the law. Dormer evidently believed the whole story.

James McParlan, alias James McKenna (z)

From Pottsville, McParlan made a series of trips through the Schuylkill field to gather more information. He described Mahanoy City, literally divided into an armed camp between the Welsh and Irish, as the most "Godforsaken" place he had ever seen. Most men carried pistols. Every ethnic group had its own organization: the Welsh and Germans were the "Modocs," the skilled Kilkenny miners the "Chain Gang," and of course the "Mollies". When violence occurred, it was virtually impossible to uncover the real motive. But McParlan was looking for only one kind of violence—Molly inspired. To get information he needed to be inside the organization, if, in fact, such an organization existed.

In April 1874, he achieved his goal: he was initiated into the Shenandoah Lodge of the Ancient Order of Hibernians. The detective took the oath, and paid a three-dollar initiation fee. Within only a few months, he had been able to establish his reputation and to ingratiate himself with the top leadership, a fact of which he was justly proud. "A final triumph," he called it. Now McParlan could begin his work in earnest. (58)

Violence was on the up rise in northern Schuylkill County. The killing of George Major, in Mahanoy City on October 31, 1874, led to a greater outcry than ever against the Irish secret society. The imminent cause of death was a violent conflict between the Welsh and Irish population of Mahanoy City.

Mahanoy City was segregated along ethnic lines. Main Street divided the Irish section of town from the British section. Like most surrounding areas, Mahanoy City was the scene of sporadic fighting between rival ethnic groups.

There were two fire companies in Mahanoy City, the Humane Fire Company # 1, staffed by all Irish and the Citizens Fire Company, staffed mostly by Welshmen. George Major, Mahanoy City's mayor, was the foreman of the Citizen's.

Mahanoy City, Pa. (aa)

When a fire broke out in the center of the town just before mid-night on Saturday October 30, 1874, the two fire companies rushed to the scene and a major brawl ensued. According to the *Miners' Journal,* the purpose of the fire was to get up a fight between the two fire companies. That incident took place on the weekend of Halloween, surely no coincidence. The celebration of the Celtic Festival no doubt added to the already existing tensions, not to mention the boisterousness of payday and the Saturday night revels that typically accompanied it. Shots were soon exchanged and several people had been injured when, just after mid-night, George Major attempted to impose order by stepping out into the street and brandishing his pistol. Someone in the crowd shot Major and before he fell, he fired off two more bullets. Daniel Dougherty, a young Irishman who had been shot in the head, was arrested under the assumption that Major fired at him in retaliation.

The newspapers were exultant. Robert Ramsey, editor of the *Miners' Journal,* concluded that Dougherty had been caught red-handed and, as Dougherty was a member of the AOH, Ramsey characterized the crime as yet another out rage by the Molly Maguire's. "The Molly Maguire's must be rooted out", he insisted, "and an example should be made of Dougherty. One good wholesome hanging, gently but firmly administered, will cure a great deal of bad blood, and save a great many lives in this community". Dougherty, in imminent danger of being lynched if he remained in the vicinity of Mahanoy City and Shenandoah, was removed to Pottsville to await trial.

When the case was finally heard, in late April 1875, the prosecution produced a string of witnesses against Dougherty, but the defense produced an equal number of witnesses who testified that the murderer was one John McCann, who had since returned to Ireland.

The case ended under dramatic and unexpected circumstances when a doctor, who removed the bullet from Dougherty's head the night Major was killed, testified. The defense demonstrated the bullet could not have come from Major's pistol; therefore, Major could not have shot Dougherty and the case collapsed. Amid a rowdy demonstration by his supporters, Dougherty was acquitted and released.

Dougherty was a member of the AOH, and that was sufficient evidence of his guilt in the eyes of the prosecuting attorney. A dangerous precedent was being set here; mere membership in the AOH was taken to be a crime. The AOH itself was put on trial, and its members were held to be guilty by association. (59)

Now that Gowen had shored up his private police force and had Pinkerton's spies in place, he was ready for his next bold-faced move to terminate the union and the Mollies.

Gowen then moved to force unity among the small operators. He established the Schuylkill Coal Exchange, which bound owners in the southern field served by the Reading, and accelerated production in order to meet pubic demands in the event of a long strike. Finally, when all was ready, wages for contract miners were cut twenty percent and those of laborers ten percent.

The miners had no choice. The long strike of 1875, which lasted more than five months, resembled a war more than a labor dispute. The miners did not seek the strike. For months, their union had voiced increasing concern over the owners' preparations, and Siney had expressed the hope that outstanding grievances could be settled through arbitration.

Gowen's tactics were ruthlessly effective. Strikebreakers, protected by the security forces, were recruited and brought in by the trainload. Pinkerton agents and armed thugs attacked strikers and terrorized coal communities. Edward Coyle, a union leader and AOH head was killed, as was another union activist when a mine boss fired without warning into a crowd of strikers, a crime for which he was later acquitted saying that he acted in "self-defense". The union, infiltrated at the highest levels by spies, had its every move anticipated. It was unable to prevent provocations that resulted in violence and harsh reprisals against strikers.

As weeks and then months went by, the violence intensified. The miners struck back as best they could. Strike breakers were beaten and murdered, police were hunted down, bosses harassed and intimidated; miners who even mentioned returning to work were sent death warnings. The miners derailed trains, sabotaged machinery, burned colliery buildings, a telegraph office, and dumped coal from laden cars.

On June 3, in the sixth month of the strike, more than six-hundred miners briefly occupied Shenandoah, and then marched on the West Shenandoah Colliery to force "scabs" to close down production. A group of agents armed with rifles met them. James McParlan, one of Pinkerton's spies, tried to provoke an incident, but the miners who had no better weapons than hickory staves, withdrew. They then moved on Mahanoy City and closed down several collieries. They joined with a group from Hazleton who were also shutting down mines, and the combined strikers marched on the Little Drift Colliery. There they came up against parts of the state militia that had been sent to the region by the governor. Only the presence of the heavily armed troops prevented widespread blood shed.

During the Long Strike, hundreds of miners' families subsisted on little more than bread and water. Men, women, and children went into the woods to dig for roots and edible plants in order to survive. Yet the public reports that appeared were often distorted. A cartoon that ran in a New York daily showed a woman putting her last loaf of bread in the oven while a group of drunken men rollicked in the backyard.

With their material resources gone, and faced by the overwhelming power of the owners and the state, the miners finally conceded. The union made a last appeal to Gowen for compromise, but he adamantly refused. By mid-June, the strike was over, and those miners not blacklisted were allowed to return to work with a twenty-percent wage cut.

Gowen boasted that he had spent four million dollars to break the strike and the union, yet peace did not return to the anthracite fields. Undaunted, the miners reorganized within the AOH and carried out reprisals. Killings, assaults, and burnings continued throughout 1875. (60)

The trouble in the summer of 1875 began in Mahanoy City as part of the continued hostilities emanating from the killing of George Major the previous October. After Daniel Dougherty had been acquitted of the killing of Major in April, the Welsh "modocs" had apparently sworn revenge. Late in May, Dougherty was attacked and shot by two men. The bullet passed through his clothing and he survived the attack unscratched. John Kehoe, a prominent figure in the AOH, called a convention of the AOH to avenge the assassination attempt on Dougherty. The convention met in Mahanoy City amid scenes of violence and crowd protests in the Mahanoy Valley. Three men were blamed for the attack on Dougherty, 2 kinsmen of George Major, William and Jesse Major and their fellow Welshman, William "Bully Bill" Thomas. McParlan "the Pinkerton spy" who was present at the meeting agreed to the following plan. The Major's had left Mahanoy City and were working at a mine in Tuscarora. The AOH body-master in that region, John "Yellow Jack" Donahue, was assigned to take care of them, while the Shenandoah Division would handle "Bully Bill" Thomas. McParlan notified Thomas Hurley, John Gibbons, and Michael Doyle of the plan. These three men then apparently decided to carry out the Thomas killing the following day, but troops had just been stationed in the Mahanoy Valley, following the disturbances over the weekend, and McParlan claimed to have dissuaded the men from carrying out their plan. The return of the Shenandoah body-master, Frank McAndrew, relieved McParlan from the responsibility of planning the assassination. Hurley, Gibbons, and mineworker named John Morris finally attempted to kill Thomas in a stable near Shenandoah on the morning of June 28, 1875. Thomas was shot twice, but survived. At no point did McParlan make any attempt to warn him. Gomer James, a miner, was the next victim in the ongoing dispute between the Welsh and Irish residents of the Mahanoy Valley. James had been acquitted of the murder of AOH member, Edward Cosgrove, in Shenandoah in 1873. He was no doubt a marked man there after.

On August 14, 1875, James was tending bar at a picnic just outside Shenandoah when a man with a revolver walked up to the bar and shot him dead. Though the killing occurred in broad daylight, the assassin apparently disappeared into the crowd, and nobody was arrested.

Gomer James was assassinated on what turned out to be the single most violent day in the summer of 1875. *The Miners Journal* lamented the events of this drunken and violent payday with a banner headline:

BLOODY NIGHT NORTH OF THE MOUNTAIN

"Squire Gwyther of Girardville assassinated.
Gomer James of Shenandoah butchered at a picnic.
Fight in Mahanoy City and a man fatally shot.
"Bully Bill" in a row and comes to jail.
Man with an oyster knife in his back."

In Girardville, a fight broke out at one of the town's taverns. The victim went to the office of Thomas Gwyther, justice of the peace, and swore out a warrant for his assailant's arrest. However, when Gwyther stepped out on the porch of his office to serve the warrant, he was shot dead. A man named Thomas Love was arrested for the murder, but he soon was released on the grounds that the assassin had been his brother, William, who had fled the region.

The final incident on this "bloody Saturday", involved the Welshman "Bully Bill" Thomas, once again. This time he engaged in a shoot-out with an Irishman named James Dugan on the main street of Mahanoy City, in which an innocent German bystander was hit by a stray bullet and killed. Thomas was arrested for the assault on Dugan, but no charges were brought for the killing, presumably because Thomas decided to cooperate with the Coal and Iron police who needed evidence against the Mollies

In August 1876, nine prominent AOH leaders were tried for conspiracy to murder Thomas; but Thomas himself, despite his violent reputation, never faced charges. Many of the Irish residents of the anthracite coal region must have wondered whether there was one standard of justice for themselves and another for the Welsh. (61)

Several highly publicized murders took place during McParlans' (alias McKenna) stay in the coal region, including that of Benjamin Yost, a police officer who had crossed the Mollies by arresting and beating member Thomas Duffy in the early morning hours of July 14, 1875. Three men waited in a cemetery near the end of Yost's beat in Tamaqua. As the police officer climbed a ladder to extinguish a street lamp. Huge McGeehan and James Boyle stepped forward and shot him. Yost fell mortally wounded while his attackers, along with their guild, James "Powder Keg" Kerrigan, made their escape.

The organization sometimes asked members from one lodge to carry out assignments in another jurisdiction. The advantage being that Mollies from outside the area would not be recognized. The plot to kill Yost was allegedly hatched in the Tamaqua tavern run by James Carroll and carried out by McGeehen and Boyle of Summit Hill. After the Yost killing, two Mollies from the Laffee District, Michael J, Doyle and Edward Kelly, were commissioned to gun down Welsh mine Superintendent John P. Jones of

Tamaqua. The Mollies accused the superintendent of blacklisting miners who had taken part in the strike. On September 3, Jones was shot in the back as he walked along the pipeline that led to the Lehigh and Wilkes-Barre Coal Company mine in Lansford, Carbon County.

In a similar fashion, Thomas Sanger, foreman of Heaton's Colliery in Raven Run, near Girardville, and miner William Uren had been gunned down two days earlier as they walked along an empty street to work. Sanger died because of an alleged workplace grievance, while Uren who boarded with the Sanger family, was slain to eliminate him as a witness.

The violence, however, was not all one-sided. The most noteworthy case of the tables being turned took place in Wiggans Patch, near Mahanoy City, in the early morning of December 10, 1875. A group of armed and masked men burst into the home of three men believed to be involved in the deaths of Sanger and Uren. The vigilantes killed suspected murderer Charles O'Donnell and the pregnant wife of Charles McAllister. McAllister was wounded but survived. Moreover, in the frenzy and confusion, one of the attackers pistol-whipped McAllister's mother-in-law. The true identity of the Wiggan Patch attackers was never established, but rumors blamed the attack on a group of irate valley residents trained by Captain Linden of the Coal and Iron Police.

The Wiggans' Patch incident came as a shock to the Mollies. How had the attackers known that McAllister and O'Donnell were involved in the Sanger and Uren killings? The organization was further shaken by a series of recent arrests, indicating that there was an informer among them. On February 23, Shenandoah body-master Frank McAndrew warred McParlan (McKenna) that Kehoe was laying bets that he, McKenna, was the spy among them. Instead of fleeing for his life, however, McKenna confronted Kehoe at his bar, in Girardville, and demanded that he call a conclave of the organizations' leaders so that he could defend himself. Kehoe agreed to the meeting, but secretly assigned two men to murder McParlan (McKenna). Two weeks later, James McParlan disappeared from the region when he learned that Kehoe planned to have him killed. McParlan left the region on March 7, 1876 on an early morning train bound for Philadelphia. (62)

McParlan met with Pinkerton agency chiefs. He provided them with a list of all the known Mollies in the five anthracite counties.

Three Mollies were arrested for murder of mine boss John Jones. The three had been arrested on scant evidence. The first "Molly trial" was that of Michael Doyle in January 1876, did result in a conviction.

The final demise of the Mollies began in the Schuylkill County Courthouse in Pottsville on May 4, 1876.

Five men, James Carroll, James Boyle, Hugh McGeehan, Thomas Duffy, and James Roarity went on trail for the murder of policeman Yost. The evidence against them had apparently been provided by Molly, turned informer, James "Powder Keg" Kerrigan.

On the day of the trial, the courtroom was filled with relatives and friends of the accused. Many had journeyed since before dawn to be present. When the defendants were brought in, wives, fathers, mothers, sisters, brothers, and children rushed forward to touch and embrace them. The court allowed the unusual procedure of seating the wives next to the prisoners. Hugh McGeehan's young bride sat in an uncomprehending daze, and James Carroll's son climbed onto his father's lap. The prosecution sought to impanel as many Pennsylvania Dutch as possible, because of their well-known antagonism to the Irish. One would-be juror admitted that he knew little English and would not be able to understand the witnesses, while another admitted that he had been a member of a vigilante group. The judge, nevertheless, allowed them to be seated. It became clear that the proceedings were stacked against the Mollies.

In a trial filled with sensations, perhaps the greatest was provided by Franklin B. Gowen, President of the Philadelphia & Reading Railroad who made an appearance on the second day as, incredibly enough, the Chief Counsel for the Commonwealth. Dressed in formal evening clothes, the special prosecutor, henceforth, dominated the proceedings. He was the lead actor in a drama of his own creation. The following day, the prosecution electrified the courtroom when it announced that James McParlan would appear as the chief witness. Almost at the same moment, the news spread throughout the courtroom that eleven more Mollies, including Body-Masters John "Black Jack" Kehoe, "Muff" Lawler and Pat Butler, were being led into the courthouse in chains, victims of yet another sweep. Gowen had orchestrated these arrests to demoralize the defendants. This brilliant tactic worked.

For four days, under the careful questioning of Gowen, McParlan held the stand, providing a detailed account of his thirty months among the Mollies, a performance that he was to repeat at latter trials. His feet rested on the rail before him. The detective revealed all that he supposedly knew about the killing of Yost, including the information that all the defendants had confessed their crimes to him personally. This was probably sufficient evidence for conviction but Gowen was not satisfied. On the second day of McParlan's testimony, District Attorney Kaercher revealed the real motive of the trial, telling the jury that the detective had been in Schuylkill County to become familiar with the secret association known generally in the locality as the Molly Maguires. However, the real name of which is the "Ancient Order of Hibernians and that it was practice for members to aid and assist each other in the commission of crimes, and in defeating detection and punishment." This makes it clear that it was not merely the defendants who were on trial, but an entire organization.

James "Powder Keg" Kerrigan (bb)

The second key prosecution witness was, James "Powder Keg" Kerrigan, the four-foot-eleven-inch tough, hero of the Civil War, father of fourteen children, and participant in at least two murders. Again, under the guidance of Gowen, Kerrigan gave damaging testimony and at the same time playing down his role in the crimes, he described. His wife, called as a defense witness, repudiated his testimony, calling him "a little rat". But Gowen skillfully made her admit that she was really angry with her husband, not for his alleged crime but because he had turned informer, an offense worse than murder in the Irish community.

The defense brought forward a series of witnesses who of course contradicted McParlan and Kerrigan's testimony. When John Mulhean, one such witness, was undergoing cross-examination, Gowen suddenly demanded that he be arrested for perjury on the basis of what he knew from McParlans' reports. The court was thrown into an uproar. The point was further driven home during the cross-examination of another defense witness, when Gowen launched into a legally questionable diatribe, "Enough has been proved in this case, in this court room, in the last ten days to convict of murder in the first degree every member of that organization in the county, for every murder committed in it. Every member of that organization is in the eyes of the law guilty of every murder as an accessory before the fact and liable to be convicted and hanged by the neck until dead." When the jury returned its verdict, few observers were surprised. All the defendants were convicted. There were other Molly trials, but the Yost case was the most decisive one. Gowen achieved his purpose. In stacked trials, the Mollies were destroyed.

Perhaps the real meaning of the Molly trials came out during the judgment of Thomas Munley, accused of the murder of mine boss, Thomas Sanger. When a witness was asked to identify Munley, he said, "That is not the man I recognized at all!" Munley's lawyer jumped to his feet and pleaded to the jury. "For Gods sake, give labor an equal chance. Do not crush it. Let it not perish under the imperial mandates of capitalism in a free country!" Munley was convicted and hung anyway. (63)

Pennsylvania's Day of the Rope, June 21, 1877. Pottsville Pa. (cc)

June 21, 1877, would be a sad day for Irish Americans who came to this great land called America hoping to better their lives, only to find a hangman. That day would be remembered as "Black Thursday—Pennsylvania's Day of the Rope." Ten Irish American's were hung by the neck 'till dead.

Alexander Campbell

Newspaper sketch of the
hanging of Thomas Fisher
inside the Mauch Chunk Jail
(The Old Jail Museum)

The Jail at Mauch Chunk

Newspaper artist's sketch of scene
outside the Mauch Chunk Jail,
before hangings.
The Daily Graphic, New York
June 22, 1877
(dd)

In Mauch Chuck, at the Carbon County prison, wives and families
huddled on the steps waiting for word that four men were dead. They had
said their good-bys the night before in their cellblock. Three-hundred
people, including reporters from the *New York Times* and the *Philadelphia
Inquirer* were gathered to watch the execution of four members of the Molly

Maguire's convicted of murdering two coal mine bosses. Even though the sun poured in through the skylight and warmed the slate block floor, there was a chill of death in the air as Alexander Campbell, John Donohue, Michael Doyle and Edward Kelly, shackled in chains, walked to the gallows, specially constructed to accommodate four people for the purpose of ending their lives at the same split second.

Campbell was the first to climb the gallows. He, Kelly and Doyle were convicted of the 1875 murder of mine boss John P. Jones of Lansford. Doyle was next, followed by "Yellow Jack" Donohue, who was found guilty of the murder of Summit Hill mine boss Morgan Powell in 1872.

The priest asked the four men to kneel and all were given absolution. After the priest left the platform, the sheriff and his deputy removed the chains and slipped ropes around their necks.

Campbell proclaimed his innocence to the last moments of his life. Before being led out of his cell to the gallows, he placed his hand on the wall of his cell leaving his handprint, declaring its imprint would remain as a sign of his innocence. To this day the handprint remains, even after numerous attempts to remove it. (64)

In near by Pottsville, the seat of Schuylkill County, a similar drama was being enacted. While an emotional crowd of about two-thousand family and friends were clinging to the hills surrounding the prison yard six more Irishmen went to their deaths. All marched to the gallows stoically, seemingly indifferent to their fates, but to the end protesting that they were not guilty. The first to mount the double gallows were Huge McGeehan and James Boyle; McGeehan clutching a crucifix in one hand and a small statue of the Blessed Virgin Mary in the other, while Boyle held a blood red rose. Before the trap was sprung, Boyle shouted to McGeehan, "Good bye old fellow, we'll die like men!" Then the hangman summoned James Carroll, father of four children, and James Roarity, a recent immigrant, who had received a letter from his father in Ireland just days before, declaring his belief in his son's innocence and asking him to trust in God's mercy. Finally, Thomas Munley and Thomas Duffy died, despite the protests of a Roman Catholic priest, Father Daniel McDermott, who swore that Duffy was innocent.

Within the next year and a half, ten more Irishmen died on the gallows, convicted, like those before them in controversial and seemingly unfair trials of murder and other crimes against mine company bosses and local law officials. (65)

In the years since the accused Mollies went on trial, opinion about the organization has been divided. At the conclusion of the court proceedings writer F. D. Dewees summed up the era of the Molly Maguires as, "A reign of blood—they held communities terror bound and wantonly defied the law, destroyed property and sported human life." Other writers have characterized the Mollies as more sinned against than sinners, the victims of mine owners bent on destroying the labor movement. The trials were held at a time of strong anti-Irish prejudice and were often preceded by prejudicial newspaper accounts. Not a single Irish American was on any of the juries. Sympathetic judges allowed Gowen to rant on endlessly about the Mollies, often painting an even more sinister picture than the facts supported. As Carbon County Judge, John P. Lavelle noted in his 1994 book, *The Hard Coal Docket*, "Any objective study of the entire record of these cases must conclude that they (the Molly Maguires) did not have fair and impartial juries. They were, therefore, denied one of the fundamental rights guaranteed to all of Pennsylvania's citizens."

All told, twenty men were found guilty of murder and sentenced to be hung by the neck until dead. Some of the Mollies were probably innocent of the crimes for which they were accused. One of the more questionable convictions was that of Alexander Campbell, who was charged with masterminding the slayings of mine Superintendent Morgan Powell in 1872 and John P. Jones in 1875. Campbell was never proven to be connected with the actual perpetration of any Molly crimes, but the testimony of "Powder Keg" Kerrigan, who turned states evidence and escaped punishment, sent Campbell to the gallows.

John "Black Jack" Kehoe was hanged in 1878 for the murder of mine foreman Frank W. Langdon. A century later, Pennsylvania Governor Milton J. Shapp granted Kehoe a posthumous pardon. The Governor wrote, "It is impossible for us to imagine the plight of the nineteenth century coal miners in the anthracite region of Pennsylvania, and it was Kehoe's popularity among the miners that led Gowen to fear, despise, and ultimately destroy him." Shapp continued, "We can be proud of the men known as the "Molly Maguires", because they defiantly faced allegations which attempted to make trade unionism a criminal conspiracy." (69)

This poem was found in the jail cell of Michael Doyle after his hanging on June 21, 1877. It appeared in the *Mauch Chunk Gazette* July 14 1877.

> Prisoners as long as we are innocent,
> And our conscience light and free.
> We will trust in God and be content
> With whatever His will may be.
>
> But still alas! I cannot rest,
> While my parents they do sigh,
> I pray that God will them bless,
> And grant them heaven when they die.

Ever since we being imprisoned,
It has caused their hearts to grief,
And they did their whole endeavors
To obtain their son's relief.

Grant heavenly Father, the star of light,
Pray comfort them each day,
And lift their souls then upright
When they are called away.

Who could find better parents than mine,
Who proved themselves to be.
Who stood to me through thick and thin
To regain my liberty

It has pierced my feelings so severe,
And the heart starts in my breast
To think that I am the cause
Of their troubles and distress. (70)

Chapter Seven

SOME OF THE FUN TIMES

Though life in the mines was hard and monotonous, and vacations from the long, six-day workweek were unheard of, except for periodic layoffs and strikes, there were various diversions to fill the miners' ideal time. Certainly one of the most important diversions was the tavern, and there were many of them.

The Drinking Capital of America

Mahanoy City   1903

Population: 13,725 — A total of 143 barrooms, or for one for every 98 inhabitants.

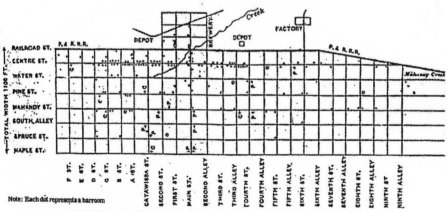

RUNNER-UP: Shenandoah (Population 20,700 — A total of 174 barrooms, or an average of one for every 119 inhabitants)

86

This chart says it all about the rip-roaring days when the Mahanoy Valley was the home of the hardest-working, hardest-drinking bunch of men in the history of America. At the time, the chart was made, the violent 1902 strike, which brought troops to the valley, had just concluded. After suffering through six hot and dry months, there was no welfare or unemployment compensation in those days, the mines were back on full time, the miners' pockets were full, and so obviously, where many miners. Those were the days when the local folk had the best beer you could find, real tonic brewed right in the old home town by Kaier's Brewery in Mahanoy City and by the Columbia and Home Breweries in Shenandoah. Here's to those good old days!

(kk)

Among the Irish, accustomed to hard drinking in their native land, alcoholism became a problem. Even contemporary observers were surprised by the amount of alcohol consumed in the coal region. Until the 1850s, alcohol was sold at many collieries. But pressure from the miner's wives and the realization by the coal operators that drinking on the job affected production resulted in prohibition in the mines.

The taverns and porches of general stores were the scenes of much tough and raucous entertainment. There the miners regaled each other with stories, tales, ballads, and songs that expressed their thoughts and feelings, wit and humor, hopes and fears.

Favorite visitors to the taverns and patch's were the roving minstrels who wondered through the coal region singing ballads, telling stories and dancing. Though often profligate and irresponsible, they were welcome at every social gathering. Tavern keepers amply supplied them with drink because of their drawing power. No wedding, christening, or wake would be complete without them. Strumming banjos or playing fiddles, the minstrels sang about events, like the great Avondale mine disaster, the Molly Maguires, or simply, the woes of working down below. Often they improvised ballads and stories about local happenings. Coins tossed into a hat were their pay. Their verses and music exuded sweat and blood, echoed every colliery sound, every colliery smell.

The Irish miners loved competitive sports, especially those in which they could display their strength and gamble on the outcome. Sizable purses went to those who could run the fastest, lift the heaviest weight, jump the furthest, or beat any man in the tavern—bare knuckles, fight to the finish.

Certain days had special significance, June 25, was celebrated by the Irish, and others from the British Isles, as mid-summers day. Old flour or cracker barrels filled with excelsior and other flammable materials were stacked around a tall pole and ignited after twilight. Singing and dancing went on around the huge bonfires that lit up the countryside. Single girls were supposed to see the faces of their husbands to be in the crackling flames.

On Saint Patrick's Day, no Irishman would work. After attending church, they paraded through town behind the church's green banner, which was a symbol of their pride. In the evening, they held a banquet, which was followed by speechmaking, singing, and humorous stories about foibles of their people. The Irish also turned July 4, into a great holiday of their own. Since it was the celebration of America's independence from England, the Irish joined in with special glee, parading in their best dress with every available member of the community.

In the evenings, miners and their families gathered in the patch's, against the backdrop of colliery buildings and culm banks, to sing, to listen to storytellers, and play folk games. A sheet of iron borrowed from the colliery provided an improvised dance floor for reels, and jigs, as the fiddler scraped out a tune. Even underground, miners entertained themselves. On their breaks, they congregated at the turnout, and danced on an old door formerly used to control the ventilation, in a vault lit only by the pale yellow flames sputtering from their teapot shaped lamps.

Catholic holy days, weddings, and wakes also provided diversions. Wakes and weddings were fine excuses for drinking and rollicking, which sometimes lasted for days and went on in shifts, to accommodate the work rhythm of the mines. (83)

The Irish did not have a monopoly on using their ethnic customs as a diversion from the grueling work in the mines.

Protestants were made increasingly uncomfortable by the appearance of more and more Catholic churches and the Sunday spectacle of hundreds of Catholics marching in procession in the streets behind a priest carrying a seven-foot high cross. One old Baptist Deacon, who visited a Slovak section in Mahanoy City on a Sunday afternoon exclaimed, "It was terrible. Saloons full blast, singing, dancing and drinking everywhere. It was Sodom and Gomorrah revived. The judgment of God, Sir, will fall upon us." (84)

Slavs were heartily criticized for their ceremonies, which were not understood by outsiders, who often considered them demonstrations of a barbarous and corrupt culture. Since many of their ceremonies involved drinking and dancing, conservative Protestants eyed them with particular suspicion.

One wide-eyed reporter, for a national publication, who witnessed a Hungarian baptism celebration, wrote about it with a mixture of wonderment and displeasure, "After leaving the church," he reported, "the party returned to the house, where the host filled a huge vat with beer and laced it with two jugs of whiskey and a handful of hot peppers. While the mixture called *polinki* was being stirred, the Hungarians sang and danced in a circle around it, like Apaches, first on one foot, then the other. During the course of the celebration, as things heated up, the newly baptized infant was removed from the house, safe from harm." (85)

The mining towns provided limited entertainment for young people. There were occasional dances, and other social activities, sponsored by the churches and fraternal groups, usually under the watchful eyes of the parish priest and adults. For adolescent boys and young men who had or took more freedom, there were more questionable diversions in those roughhouse mining towns, there were all kinds of gambling. The priest and ministers deplored the tendency of the young to squander their money on such activities. Drinking was learned early, and the time from fetching a pail of beer for "the old man," to taking a first nip at the bar was short. (86)

The noon recreation hour, Kingston coal company.
Courtesy of the Wyoming Historical and Geological Society, Wilkes-Barre, Pa.

(11)

These mineworkers played as hard as they worked. On a Sunday afternoon, in the fall, you would find these young men participating in their favorite sport . . . football. Football was a good way to expel the tensions built up during the grueling workweek, while earning the bragging rights for your patch or town

Each colliery had its employees represented with a team that played in a league. The games were well represented with family and friends as spectators, and these loyal fans weren't afraid to bet on the outcome.

Football in the coal region was of that good of caliber, that in 1925, Pottsville got a team together and entered the National Football League.

It all began in 1924. Dr. J. G. Striegel called Joseph Zacko, owner of a local sporting goods store, and placed an order for twenty-five matching jerseys. The color wasn't important. When Zacko delivered twenty-five maroon colored jerseys, the Pottsville Maroons were born.

Dr. Striegel believed his team was equal to any team in the world. He attended the National Football League meeting in August 1925, and paid a five hundred dollar application fee, posted a twelve hundred dollar guarantee, and received a franchise for Pottsville in the National Football League. (87)

I can remember the old-timers bragging about how great the Maroons were. There is no doubt that the Maroons were good. I would hear how they complied a 10-2 record their first year in the league. I would hear how, on a cold and snowy day, in 1925, they went out to Chicago and beat the Cardinals and were crowned the NFL Champions. I would hear how they beat the Fighting Irish of Notre Dame in an exhibition game. This was the same Irish team that won the National Championship in 1924 and had the famed "Four Horsemen" playing in the game. All this was true, but there's more to the story. It seams that the NFL did not approve the exhibition game against Notre Dame and this resulted in the NFL suspending the Maroons from the league. This decision was controversial because the Maroons said they were given verbal permission to go ahead with the game. To this day, the controversy goes on.

In 1926 the Maroons were reinstated into the league and compiled a pretty good 10-2-1 record.

In 1927, Pottsville lost several of its stars, and others were growing older. The team slipped to 5-8 that year, then to 2-8 in 1928. Dr. Striegel sold the franchise to an ownership group in Boston, where they played as the Bulldogs in 1929. Late that season, the Boston Bulldogs played a home game in Pottsville as a farewell to the fans. And, just a little later, the Bulldogs bade farewell to the NFL after compiling a modest 4-4 record and even more modest attendance figures.

Because the Washington Redskins began in 1932 as the Boston Braves, some Pottsville backers, with the help from a few writers, have suggested that the Redskins descended from the Maroons by way of the Boston Bulldogs. But it's not true. The 1932 Boston franchise had no relationship to the 1929 Bulldogs. (89)

This could be a mere romanticism, but when I see the Washington Redskins take the field on a Sunday afternoon, I can't help but to notice the maroon colored helmets and uniforms worn by this modern day NFL team!

The toughness inherited by the sons and grandsons of those immigrant coal miners made the caliber of sports in the coal region a benchmark for others to follow. Today whether it is football, basketball or baseball it's a pretty good bet that there'll be a team from the hard coal region competing in the state finials.

Some of these gifted athletes were spared the dilemma of following their father's footsteps into the mines by receiving a calling to play sports at a professional level. I could dig into the record books and brag about some of the achievements of some of these "pros", but the one I admire the most was a ballplayer by the name of Pete Gray. Pete's Major League career spanned only one year and his life-time batting average was a mere .218. However, the story of Pete Gray goes far beyond a short stay in the Majors.

Pete was born in the coal-mining town of Nanticoke, Pennsylvania. At the age of six, Pete lost his right arm in a truck accident, but Pete never

Pete Gray (mm)

lost his love for the game of baseball. Despite being naturally right-handed, he learned to hit and throw as a southpaw, and eventually rose from the semi-pro leagues in the coal region, all the way to the majors.

In the field, Pete wore a glove without any padding. When the ball was hit to him, he made the catch with the glove directly in front of him, normally about shoulder high. As the ball hit the glove, he would roll the glove and ball across his chest from left to right, somehow in the process; he learned to separate the ball from the glove. In the same motion, the glove would come to rest under the stump of his right arm and the ball would end up in his left hand.

In playing ground balls, he would let the ball bounce off his glove and grab the ball while it was still in the air. Some observers said that this process allowed Pete to field ground balls faster than other fielders that didn't face the same handicap as Pete. When he was backing up another out fielder, he would drop his glove and be ready to make the play barehanded.

At the plate, despite having just one arm, Pete used a full weight, regulation bat. He was described as standing back behind the plate and cocking the bat as any other hitter would. His hand was about six-inches up the bat handle and he would take a full swing.

Pete played for Three Rivers of the Canadian-American League in 1942 and hit .381 in forty-two games.

He moved to Memphis in 1943 and played center field, and hit .289. In 1944, he put together a season that would get him noticed by the Major League scouts. He hit .333 with five home runs, and stole sixty-eight bases to earn the leagues Most Valuable Player Award.

The St. Louis Browns, of the American League, signed Gray in 1945. He made his Major League debut on April 18, 1945 and collected a hit in four at bats. On May 20, 1945, Pete had one of his best days in the Majors against the New York Yankees, at Yankee Stadium. The Browns beat the Yankees 10-1 and 5-2 in a doubleheader. Pete had two runs batted in on three hits in the first game and a hit and scored the winning run in the second game.

Gray played in seventy-seven games and had two hundred thirty-four at-bats. He had fifty-one hits, thirteen runs batted in and five stolen bases. (90)

Pete was sent down to the minor's after the 1945 season. He played for various minor league teams 'till he retired in the 1950s. Pete returned to Nanticoke, where he lived until it was his turn to climb those marble steps to the ballpark in the sky.

This coal cracker's remarkable achievements over his lifetime are certainly an inspiration to someone who is physically challenged. It's a pretty good bet that Pete Gray never considered himself handicapped.

Pete was without a right arm; but he wasn't without courage.

Although many of these coal crackers didn't have much schooling, some of the lessons learned from their living experiences in and about the mines couldn't be taught in any books. They learned the value of life. Each day they would enter the mines not knowing if they would return home to their loved ones. If by fate, the *whistle* didn't blow and the men returned home safely it would be cause for much jubilation. They learned to celebrate life. Be it work or play, they lived life to the max, because just maybe tomorrow the *whistle* would blow and a wife would become a widow and a child would be without a father.

Chapter Eight

1963 CAVE-IN AT SHEPPTON

On Sunday night, 42 year-old Louis Bova climbed out of the slope of the Fellin Mining Company coal mine near Sheppton, in Schuylkill County, and went home to his wife and eight-month old son in nearby Pattersonville.

"Eva," he said, "that mine is a dangerous place, I can hear the creaks." All the same, Bova went back into the mine on Monday morning. Soberly he went, deep into the earth, trying to stifle his misgivings. He could not do otherwise; this was his bread and butter.

When he came out at the end of his stint he told the mine owners that he wanted to take the afternoon off so he could get home and spend some time with the baby. He said that he dotted on the child all the more since it was a boy.

He did not know that it would be Gods' will that this would be the last time he would be with his wife and child. Bova was instructed to report for work Tuesday morning.

August 13, 1963, Bova, Henry Throne, and David Fellin reported for work at the Fellin Coal Company. It was a nice, sunny, Tuesday morning when the trio entered the pit at about a quarter past seven. They got down more than three-hundred-feet by seven thirty and by eight o'clock, they sent a load of coal up to the surface.

Joseph Walker from Mahanoy City, who was the hoisting engineer at the operation, noticed the first sign that something was wrong. He said that, "About ten minutes before nine in the morning, he had just completed hoisting a load of coal from the slope and was returning the empty "buggy" to the mine face four-hundred-fifty-feet below ground." (The "buggy" refers to a small gondola, which runs on a narrow gauge railroad track and is pulled to the surface by a power winch.) Walker said, "That as the empty buggy was descending, he felt it come to a stop through his operating controls and knew that it had hit an obstruction." What had happened was the walls of the mine caved in and filled it with coal and rock blocking the only escape route.

Help was summoned immediately and by early afternoon, four federal and state mine officials entered the mine in an effort to determine the extent of the cave-in. William Witting said, "he was familiar with the slope and that he had gone in as far as he could and found at least one-hundred-feet of the slope was choked with running coal, and it was still running."

Gordon Smith from Pottsville, a Deputy Secretary of Mines was in charge of the recovery operation.

Throughout the afternoon Smith and state inspector, Pete Nino made several sorties down the slope with safety lamps to test for black damp gas emanating from the mine. The deadly black damp was pushed out of the cavity by the rushing of coal hundreds of feet below.

On the first try they descended about one-hundred-feet before black damp extinguished their safety lamps. They returned to the surface and their crews rigged additional air pumps to try to clear the air.

A half-hour later, they again descended the slope and this time the gas halted them at a mere fifty-feet. On a third try, they were able to descend about seventy-feet.

At four o'clock in the afternoon, the trio descended between one-hundred-fifty and two-hundred-feet below a "knuckle" of the steep eighty-two-degree slope before they had to retreat. Smith said that he could hear coal and rock still falling and that the crews would not be able to do anything, even if the gas was cleared, because of the danger of the rush of coal continuing. Mine officials at the scene indicated that there was no hope the three miners would be found alive. H. B. Charmbury, State Secretary of Mines, suspended rescue attempts Tuesday night after rescue workers were hampered throughout the day by poor air quality and the possibility of another cave in.

Day 2

On Wednesday, August 14, mine inspectors made hourly checks of the ill fated mine and reported that additional coal falls took place during the late night hours.

Deputy mine secretary, Gordon Smith said, "At nine o'clock Wednesday night everything was quiet but about a quarter past eleven, the fall started and continued until eleven o'clock the next morning when once again it became relatively quiet, although the timbers in the mine were creaking from the tremendous weight against them."

Day 3

Smith said, "We would continue to watch it and examine it carefully until such time we feel it is safe to work. We're going to wait and check it at regular intervals in the hope that the fall will stabilize itself so that we can enter the slope."

Rescuers debated on a plan to dig another shaft in an attempt to reach the three miners. The plan would be to fill up the present shaft; then dig another to the point where the miners were believed to be trapped.

Mine officials outlined the plan, which would take about fifty days, as a way to cope with the constant danger of more rock falls in the present shaft. Besides that, the rescue efforts had been slowed by the threat of deadly gases. Officials made periodic attempts to descend but had come nowhere near the three-hundred-eighty to four-hundred-fifty-foot level where the three were believed to be. This plan had to be scratched because of certain objections that were raised by the families of the entombed miners. For the families to agree to this plan would be to give up hope. (Certainly, no man can live without food or water for fifty days.)

This was a small independent mine consisting of two shifts with four men per-shift, three miners and a hoisting engineer. This mine was not a member of the United Mine Workers of America.

Day 4

On Friday, August 16, relatives of the entombed miners came to the Hazleton office of the United Mine Workers and urged the officials to intervene in the situation. The families of the men were gravely concerned that an agreement had been reached between the Pennsylvania Department of Mines and the Federal Bureau of Mines that further rescue efforts were futile and the miners should be declared dead.

Upon hearing this allocation, the office sent UMWA representatives to the scene of the disaster to do everything possible to start a rescue operation.

The UMWA released this statement; "Although the United Mine Workers of America has no contractual responsibility at this mine, our union is always concerned of all coal miners and their families. In conclusion, the United Mine Workers of America intends to pursue the course of action which it has initiated and will not relent in its efforts until the men in these so-called small operations are given the same protection as all other citizens employed in the American coal mining industry."

Joseph Fellin, brother of one of the entombed miners and a miner himself, met with the UMWA officials. Fellin stated, "That on the basis of reliable information from persons familiar with the mine, the three miners might have escaped the cave in by reaching the shelter of a gangway sixty-feet up from the bottom of the slope, and it (the gangway) runs east and west for about one-hundred-twenty-five-feet."

Fellin asked, "That boreholes be drilled into the gangway, one for intake air and another for exhaust".

The UMWA and the Pennsylvania Department of Mines agreed on the proposal. Now, instead of a recovery operation it will become a rescue operation

Day 5

A drilling unit was put in place and was ready to start drilling operations at 6:30 p.m. Saturday, August 17. After drilling for twenty-two hours, a six-inch hole broke through the gangway where rescuers hoped the three miners managed to get to.

Day 6

John Biros, of Sheppton and friend of Fellin, shouted into the hole, hoping for a reply, seconds later Biros leaped back from the hole and exclaimed, "They're alive! I hear them! They're alive!" Fellin was heard saying, "Who are you? We're all okay!"

They lowered a light down through the six-inch hole that was drilled into the chamber. Gene Gibbons, co-owner of the mine, hollered down to Fellin, "Can you see the light?"

Fellin responded, "No!"

Gibbons asked, "Where are you?"

Fellin responded, "Behind some timber, near a pipe! We're blocked in a hole and can't get out!"

Day 7

After a couple of hours went by; Fellin was able to make his way to the location where the light was lowered and was heard for the first time over a microphone that was dropped down the hole on a cord and hooked up the a speaker on the surface.

Fellin asked, "Send us down some soup, and send Hank a stogie (type of cigar)?"

Fellin was asked about the other two men, he replied, They're asleep, we're not afraid of anything down here. Do you want to know what sulfur water tastes like? What day is it?"

One of the rescue workers replied, "What day do you think it is?"

Fellin: "Monday."

Rescuer: "You're right."

Fellin: "My buddy Henry thought it was Tuesday. Tell Anna (Fellin's wife) I was asking about her and tell her to make a pot of garlic soup!"

Rescuer: "Don't sleep until we get some food down to you!"

When asked about the third trapped miner, Louis Bova, Fellins' reply was, "He's okay, but he's hurt a little bit."

Upon hearing, the news of his father surviving the cave in, Joseph, David Fellin's only son, rushed home to tell his mother the news, running in the door he exclaimed, "Mom, Mom, Pop's alive!" His mother responded, "Oh, God is so good."

The next step was to bore a twenty-four-inch hole so that they could bring the miners up to the surface. Ralph Ditzler, district mine inspector, said it would take two or three days to complete the job. Ditzler wanted to start the rescue drilling but had to hold up because the trapped miners were too tired to direct the operation. "I tried to get Fellin to tell us where to drill, advising him we could drill while they slept," but Fellin said, "No, no, we've got to get some sleep, then you can start."

So all activity outside the mine stopped and silence fell over the area as the miners slept peacefully some three-hundred-thirty-feet below. Rescue workers knelt down and prayed.

When Fellin awoke, he reported that Bova was about twenty-five-feet from him and Throne. He said he talked to Bova at intervals and that Bova apparently suffered a hip injury. Later he lost contact with Bova and thought perhaps he went to sleep or changed location. Fellin and Throne tried to clear debris to reach him but were unsuccessful. Throughout the day and night, Fellin could be heard calling "Lou! Lou!"

Rescuers began drilling a separate lifeline hole down to where Bova was believed to be, but it had to be abandoned when the six-inch hole "drifted". Within an hour, they started a new hole at a new location and got down ninety-feet before they had to stop. Fellin and Throne cautioned them to drill slower to prevent the drill from drifting.

The jubilation that had been there Sunday night and early Monday was gone. In its place was a grim determination to get the men out safely. Relatives and friends continued their vigil. Mrs. Fellin was overheard saying, "If you don't believe in God, go through something like this then you'll know there is a God."

Rescue workers were busy lowering food and other necessities down to the trapped miners. There still was no word from Bova since nine thirty Monday morning.

Near mid-night, there was a discussion on whether they should enlarge the first six-inch hole to twelve-inches. This move was ruled out because of the danger of dust falling through the six-inch hole and suffocating the miners.

Day 8

At 1:00 a.m. Tuesday, August 20, the rescue workers, on the advice of Fellin, were preparing to drive a twelve-inch hole about five-feet west of the first six-inch hole. This operation calls for the twelve-inch hole to be expanded to twenty-four-inches. If all goes well, the miners could be brought up to the surface through the twenty-four-inch hole. The drilling rig was moved into position and began drilling at 2:45 a.m. After drilling twenty-two-hours, to a depth of one-hundred-ninety-three-feet, it was decided to halt the drilling operation on the advice of the two trapped miners. They said it was cracking the roof and they feared another cave in.

After conferring with the entombed miners, H. B. Charmbury said they would start, as soon as possible, another hole eighteen-feet west of the abandoned one.

Meanwhile rescuers were successful in their second attempt to drill a six-inch hole down to the location where they hoped Bova was. It could not be determined where Bova was in relation to the second lifeline hole. Bova had no food or water for almost nine days and there was no response from him.

Day 9

Drilling of the second twelve-inch escape hole to Fellin and Throne began just after dawn on Wednesday, August 21. A dense fog shrouded the area. It was so thick that it was impossible to see the top of the ten-story-tall drilling rig. Fellin and Throne were nonchalant and joking as the drill bore down.

Day 10

After thirty-one-hours, counting a six-hour delay when the drills drive shaft broke, the second twelve-inch-hole missed the chamber where the two miners were trapped. It was a staggering blow to the rescue operation.

Without hesitation a third hole was ordered. The third hole, which was expected to take at least nineteen hours, was ordered four-feet east and eight-feet north of the second hole.

Drilling hole number three commenced at 6:00 p.m., Thursday August 22, ten days after the cave in.

The two miners, when they weren't resting, kept their spirits up by singing songs that could be heard on the surface by way of the communicating system. They added to their repertoire the familiar Negro spiritual, *Massa's in the cold, cold ground.*

Day 11

At the depth of three-hundred-feet, the two trapped miners reported they could hear the drill. Upon hearing this, the rescue workers slowed down the drilling speed. Fellin and Throne were asked to keep a close watch on the ceiling above their fourteen by nine-foot chamber where they were trapped.

As the twelve-inch bit bored an escape hole through the final eight-feet into the gangway where the miners were located, rescuers remained in contact with the men below.

> Driller: "Do you see it! (The drill) Do you see it?"
> Fellin: "We haven't a glimpse yet! Can you go very, very slow?"
> Driller: "You bet your life. Dave, how's it look?"
> Fellin: "It's coming right over our heads!"
> Fellin: "Stop! Stop!"
> Gordon Smith: "How about it Dave, a little more?"
> Fellin: "Yes, a little more!"
> Smith: "Here it goes!"
> Fellin: "There's a slight crack!"
> Smith: "Any dust?"
> Fellin: "No, but we can feel the air. I can see the drill! Hold it! That's enough! It's through! Take it up!" Fellin shouted dramatically as the giant drill rig churned through to complete the escape hole at 6:23 p.m.

Smith commanded Mike Rank, the drill operator, "Take it up Mike! Take it all the way up!"

Workers jumped into the air, cheered and laughed, and then they ran the American flag up to the top of the one-hundred-ten-foot, sixty-five-ton drilling rig.

Watching from about two-hundred-yards away, the relatives of Throne and Fellin cried tears of joy while Bova's wife bowed her head, holding back tears.

There still remained the equally dangerous task of widening the twelve-inch hole to a size through which the miners could be pulled to safety.

Despite the optimistic outlook for Fellin and Throne, there was little jubilation as thoughts turned to the third trapped miner, Louis Bova, who was separated from the other two. Few held little hope for his survival.

Reaming the hole larger differs somewhat from ordinary drilling. In drilling a hole from scratch, the earth cut away by the drill bit has nowhere to go but up and is easily removed out of the way with a dozer. In enlarging

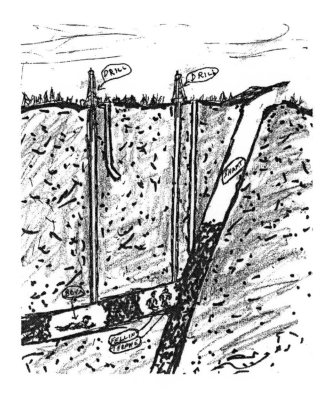

(jj)

an already existing hole, a plug is inserted into the hole about twenty-five-feet above the chamber, preventing the tons of drilling waste from filling the chamber where the trapped miners were located. Once the reamer makes contact with the plug, it will be destroyed and the final twenty-five-feet of drilling debris will fall directly into the chamber. Before the plug was inserted, rescue workers sent down larger items such as blankets, lumber, and tools that would not fit down the six-inch hole.

Day 12

As the reaming began, the surface was in constant contact with the miners below.

> Rescue worker: "Hey Davey, are you busy down there?"
> Fellin: "Yeah, we're working pretty hard."
> Rescue worker: "What are you doing?"
> Fellin: "We're putting up shoring." (Timber to support the roof)
> Rescue worker: "Are you getting any dust?"

Fellin: "Just a little. You know I would have become a priest, twenty
    years ago, in order not to go through an experience like this.
    I guess I'll get hell from the woman when I get home."
Rescue worker: "We all deserve to get hell sometime."
Fellin: "Did the strip mine bill get signed into law?"
Rescue worker: "I don't know. I think it's still in the Senate."
Fellin: "How's it going up there? We're keeping our fingers crossed."
Rescue worker: "Make the sign of the cross."
Felling: "Yeah, I got two medals from the Pope."
Rescue worker: "Everybody is praying for all of you. Get some
    sleep Davey, so you're prepared for the time ahead."
Fellin: "Not on your life; I'm playing poker down here with Hank
    and I'm not quitting 'till I get even."

The huge drill kept driving at a steady pace through the rock. Nobody could guess on the time David Fellin and Henry Throne would be released from their imprisonment some three-hundred-twenty-feet underground.

Day 13

While the drill drove toward Fellin and Throne, a three-inch drill was grinding another hole into the area where Louis Bova was believed to be, although hope was slim, as he was without food or water for thirteen days.

The rescuers knew they were getting close when they heard Fellin shout into the microphone, "Something's coming down!" asked what it was, he said, "Some clay."

"Good! That's a good sign!" Gordon Smith answered, "It means we have drilled through the plug." At this point, the drill was slowed to a crawl.

Once the escape hole enlargement was completed, the men would be brought up to the surface, one at a time, using a specially designed steel capsule.

After the drill broke through into the chamber and the reamer was raised out of the three-hundred-eight-foot deep hole, it was decided to enlarge the hole using a thirty-one inch drill bit to the depth of thirty-four feet. Into this enlarged section would be lowered a thirty-inch outside diameter by twenty-four inch inside diameter steel casing thirty-four feet long to solidify the softer earth at that depth. This caused an additional thirteen-hour delay.

Fellin and Throne decided they were going to have a party to celebrate their last night in the mine. Throne asked for some fig bars, graham crackers and some sour balls. Fellin wanted hot peppers and bologna for his dinner.

For the families of the miners watching the operation, excitement was running high. "We'll even feel better tomorrow when we can hold their hands and I'll be getting the best bartender back into my tavern." said Mrs. Throne, as she wiped away tears. Throne managed a tavern at night.

"Oh! I'm so happy, so happy!" cried Mrs. Fellin. "I'm still shaking. It's one week and six and a half days. He said he'd be up by Sunday, but I'm not going to holler at him for being a day late."

Day 14

At 12:30 a.m., Tuesday, the winch resembling a giant grasshopper, was moved into position.

It was decided at the last minute to replace the rescue capsule with safety harnesses to bring the two miners up to the surface.

At 12:56 a.m., two harnesses, two helmets, and a bucket of grease were lowered down to the miners. They greased each other up and helped each other into the harnesses. The helmets were wired for sound so that the miners could keep in contact with their rescuers on the way up to the surface.

The minutes ticked by slowly until Throne reported he was greased up, suited up, and ready to come up. It was 1:50 a.m., when the power winch started to slowly pull Throne up the rescue shaft.

"I'm coming up!" he yelled as he was being lifted, ever so gently, up the shaft. Then Throne said, "I'm turning around a bit now . . . Still turning . . . Boy what a ride, it's like the chutie-chute at Coney Island!"

Near the top he asked, "What kind of moon was out?" When he was told there was no moon, he quipped, "No moon! Holy mackerel there Andy!"

When Throne reached the surface, the rescue workers shouted, "He's up! He's up!" There was applause, cheers, and whistles. When Throne was rushed past the crowd, to a waiting helicopter, he was rapped in a blanket and appeared exhausted. Throne reported that he had plenty of room coming up the shaft; so much that he was bouncing around and hitting the walls of the rescue hole.

Now it was Davey's turn to come up. "Keep on going! Keep on going!" Fellin shouted as he was making his trip up the rescue hole. "This is the best ride I ever had!" At one point he was singing, *She'll Be Coming Around the Mountain.* Then he asked those on top, "Do you want to hear another song?" He was told to wait until he reached the surface. Fellin was snagged in the lines about half way up, but he was quickly freed.

"She's doing nicely!" he said. "Keep coming! It's working like a clock! Happy New Year! Everything's okay! What a beautiful ride!"

Fellin also was cheered as he reached the surface. He covered his eyes with one hand and waved with the other. His face, like Throne's was grimy. He talked to several people for a few minutes. Then he too was taken to the hospital by way of a marine helicopter.

Early indications, after being examined, showed that neither miner suffered from anything serious.

The snatching of the two miners from the small chamber three-hundred-thirty-feet below the earth, where they had been trapped, for thirteen days and seventeen hours, by a rush of coal and rock, was a feat unprecedented in mining history.

It was the first time the rescue, of trapped miners, was made through a borehole of such great length and such small dimensions, according to rescue officials at the scene.

The men figuratively were raised from the dead. The very first day of their entombment, many people had all but given up hope that they could be found alive.

Dr. H. B. Charmbury, who was in charge of the rescue said, "That plans would be made within the hour to go after Bova." On the advice of Fellin, a twelve-inch hole would be drilled six-feet east of the original six-inch lifeline.

On completion of the twelve-inch hole, a super-sensitive microphone was lowered down the hole in hope of hearing breathing or any sign of life The only sound coming back was that of dripping water. Pebbles were dropped down the hole and the microphone recorded the sound of a splash. The microphone was then lowered into the original six-inch hole, and the same result occurred when pebbles were dropped in, a splash sound was heard.

It was estimated that by Sunday afternoon, September 1, the enlargement of the hole to twenty-two-inches would be completed. After completion, a man was lowered into the shaft and reported no sign of Bova.

By now, all hope of finding the third miner was gone. Louis Bova's body was never recovered. The mine was sealed. Today a marker marks the spot where Louis rests in peace some three-hundred-twenty-feet below.

While in the hospital, Davey Fellin was interviewed as to how he and Henry Throne survived the first six days before having contact with the world above them.

Davey said, "A funny thing occurred on the very first day. We weren't down the mine five minutes, that morning, when my stomach started feeling out of whack. I said; let's go out for an hour or so. But the boys persuaded me to stay and get some work done. So we stayed down at the tunnel's bottom. Louis was on one side and me and Hank were on the other.

"Louis reached up to press the buzzer to let Joe on top know we needed an empty buggy. Then it happened. Suddenly, everything was coming down, timber, coal, and rocks. We could see it because the power line was still working. The stuff was coming down between us and Louis. Then it was quiet for maybe a half a minute, then the rush started again. It went on like this, starting and stopping, for some time.

"In the first couple of minutes the light from the power line went off. My heart jumped into my mouth. I kept thinking; did Joe shut the power off up top? Otherwise, it could short and the wire might burn through and start a fire that would kill us in about three minutes down there. But it didn't burn.

"For a couple of hours we could see a little around us from the lights on our helmets. But they burned out. Our matches wouldn't burn down there. That was the end of the light for the next five and a half days.

"We stayed sitting there listening in the dark as hard as we could for more rushes. We sat there against the wall that way for about fourteen to sixteen hours, in a place six-feet long, five-feet wide and about three-feet high.

"We didn't talk much. We didn't have much to talk about. We got thirty. We heard some water in the drainage holes which was about ten-feet down in the floor of the tunnel. We tried, for about three-hours, to dig for it but got nowhere.

"Hank" I said, "Let's sit down and talk this over and see what we can do." Hank said, "Let's do something! Let's get out of here!"

"Hank was a new man, not very experienced. He was shaken, but not as much as you'd expect of a new man. I'd trust him anywhere.

"We found a hatchet with a broken handle, a crowbar, and a saw that wouldn't saw. We found an old rasp and I used it to file the saw and sharpen the hatchet. Now we were back in business. We had some tools to probe around. Now we could look for a way out. But we never found one.

"I found a four-foot long pipe laying there. I plugged up one end with electrical tape that I had in my pocket. Then I shoved a piece of cable through it; tying it at the other end. Then I jammed some rags into that end. Then I dropped the pipe into the drainage hole and when I pulled it up we had a four-foot pipe filled with water. Sure, it was sulfur water, but it was water that would keep us alive.

"The first time we tried it we spit it out. The second time it stayed down. I told Hank to drink only a quarter swallow at a time. We had to ration it.

"The worst thing, those first five days, was the cold. It went deep into our bones. Because of the cold, I slept very little. Also I wanted to keep listening for any noise, any sign of movement.

"To keep warm, Hank would sit between my legs with his back to me and I'd breathe warm air on his neck and back and we'd rock back and forth. Then we'd switch. We'd do this for maybe five or ten minutes at a time. Then we'd stop, but only for five minutes because we'd get cold again.

"We'd sleep face to face with our arms around each other. We'd sleep maybe a half-hour and then the cold would wake us and we'd start rocking again to get some circulation.

"In the first few days, I could tell by looking at my fluorescent watch, what day it was. But down there, in the dark, I got mixed up with morning and night and finally the days themselves.

"On the fourth or fifth day, we were trying to find our way out. We dug through a pile of debris for about forty-feet. We dug with our hands and that hatchet. It was clawing all the way, inch by inch. As the chunks of stuff fell, Hank would take a big lump down below. It would take him about an hour to get one lump down. We worked that way for about a day and a half.

"About this time we heard what sounded like rain water running down the drainage pipes. This was one of the worst moments. It could have flooded the mine, but it lasted only about twenty minutes. (80)

"Now you ask me about the strange things me and Hank saw down there."

"That when it was apparent that neither he nor Throne would get out of the mine alive, he had gotten angry with God and said a prayer in which he demanded that his creator at least have the decency to let him know what evil he had done in his lifetime that he was forced to die 'a thousand deaths', while facing torture inside the mine?

"A short while latter, what looked like three tiny bluish fire flies suddenly appeared in the total darkness, and soon began filling the chamber with a bluish light.

"The bluish light, that cast no shadow, had provided perfect visibility.

"The small enclosure where they were trapped expanded, enabling them to move around.

"I'm positive we saw what we saw. These things happened. I'm almost afraid to think of what might be the explanation.

"We saw this door covered in bright bluish light. It was very clear, better than sun light.

"Two men, ordinary looking men, not miners, opened the door. We could see beautiful marble steps on the other side. We saw this for some time and then we didn't see it.

"We saw other things that I can't explain. But I'm not going to tell you about them because I feel too deeply about all this. (81)

"Then suddenly, I heard Louis' voice shouting, "Hey Dave, hey Hank, I've come to take you home". We nearly died. It came right out of the darkness. But we didn't hear from Louis again.

"Then we heard a voice yelling, "Look for the light!" I started digging like mad towards the sound. After about two hours, we found the light only about six-feet away. It was hanging on a cord. At last, they reached us. Then they lowered food and water down to us.

"Now that we got the light, we decided to set up housekeeping. We made us a nice bedroom, about five-feet long, four-feet wide and five-feet high. We could almost stand up in it. This is where we slept and stored supplies.

"We stayed there about two days. One morning, I suddenly heard something move. I grabbed Hank and got him out of there just in time. Our roof caved in. So we moved to another spot. We made another bedroom, about ten-feet below the first one. It was a little larger, but we still couldn't stand.

"Then we heard that big drill. The noise sounded like a huge roar and a thousand devils. The hole broke through right between us as we were lying there watching it.

"Now I could begin to figure when we'd get out. I knew it would still be four or five days. We had a lot of work to do, like reinforcing props that were holding back thousands of tons of coal.

"By now, of course, we were eating well, with that stuff they were sending down. But a funny thing, about the first five and a half days, when we had no food, we weren't very conscious of being hungry. We ate some bark from the timbers, it didn't bother us.

"They sent down that sleeping bag, and Hank spent most of his time in there. He's a great guy, but he's awful sleepy. Somebody had to stay awake and listen for anything going wrong. I did that. That's why you heard me speaking most of the time on the "mike". For the same reason I decided that Hank would go up first.

"We kept on working until the reamer came down. Then they sent down the coveralls and harnesses. Then we greased each other up. Going up, I was concentrating on working the lines right so they wouldn't foul up and spin me around the way they did Hank.

"When I finally got up, I kept thinking about my wife, Anna; where she was, how she was bearing up and was she getting any sleep.

"When I got up the lights were blinding me. The first person I saw was my brother, Joe. He came over and we shook hands and he said a little, not much. It was the first time we had spoken in twenty years. We'll be doing a lot more talking from now on."

For David Fellin and Henry Throne the ordeal was over. Even after they were found to be alive, the ups and downs of hope and despair were enough to drive rescuers and relatives to nervous breakdowns, to say nothing of the trapped miners.

But the victims proved pluckier than most humans. It takes a long time to build up a feeling of confidence in your environment if you're a coal miner. Anyone, who has been in a coal mine for any length of time, particularly when your lamp fails, knows the feeling.

When there is no light, the blackness of a coal mine, deep in the earth, has no material substance to it. It is difficult to describe so there is no real way to measure how the entombed miners got along for so long in their tiny cubical.

As strong, though, as David Fellin and Henry Throne were, we could take the liberty of assuming how relieved they were at being able to gaze up at the long lost sky, see the landscape and the happy faces of relatives, and rescue workers who never gave up hope. (82)

One sunny summer day in August 1963, three miners went down into a pit to earn their daily bread in a coal mine. After the walls caved in and being entombed for two weeks, two of the miners were raised three-hundred-eight-feet to the surface through a rescue hole with the use of a winch, hook, and harness. The third miner also came up, but to a much greater height than the first two men did, but he used the marble steps, and had two celestial escorts.

Chapter Nine

THE FUTURE OF ANTHRACITE

For over one hundred years, anthracite coal was the nourishment America used to feed its growth. Anthracite coal took America from its infancy and transformed it into a global giant. Coal was the commodity that heated our homes, powered our trains and made our steel. Thanks to coal, America was able to stand on its own two feet without the help from our neighbors that lived on the other side of the Atlantic.

America, until the 1950s, was able to fuel its energy needs from commodities produced from within its own boarders. After the Second World War, oil became America's main energy source. As our energy needs grew faster than our domestic supply, we had to turn to foreign sources to supplement our needs.

Today over half of our countries oil is purchased from suppliers overseas. Will these sources always be reliable?

In 1973 and 1979, America experienced oil embargoes. The cutoff of our oil supply provided an important warning about our vulnerability to foreign disruptions. Those crisis's should have served as lessons about our dependence on unreliable overseas sources. Instead, our dependence on foreign sources stands at an even higher level today than it did during the embargoes that plunged America into serious recessions. Today America is vulnerable. (91)

Any disruption to the flow of the oil supply into this country would send this nations' economy into a tailspin. The health of Americas' economy depends on our ability to stand on our own two feet. How strong is a country that allows foreign nations to control over half the lifeblood needed to energize its needs? There is an answer, and it lies where it all began, in the hard coal region of northeastern Pennsylvania.

Over the years, our grandfathers and great-grandfathers left behind what some might think as blemishes on a once beautiful landscape. Although unsightly, these blemishes could be a blessing in disguise.

During the bygone years of processing anthracite coal at the hundreds of collieries scattered throughout the coal region, there was left behind hundreds of mountains of coal waste called culm. Culm is the coal dust and small bits of coal that was leftover after processing. During the era when coal was king, culm had no marketing value and was considered waste. Today, with modern technology, that unsightly culm can be used to generate electricity and it can also be transformed into transportation fuel.

Already in place is the St. Nicholas cogeneration plant, near Mahanoy City, in Schuylkill County, and down the road, in Gilberton, is a second electrical generating plant. Using modern technology, these plants are producing electricity with little threat to the environment.

Another project proposed, near the Gilberton plant, is a coal gasification operation that will take the readily assessable culm and convert it into transportation fuel

Today there are an estimated fifteen million tons of ugly coal waste littering the landscape of the anthracite coal region. This is enough feedstock to last for over twenty years without taking even a single pound of coal from the earth.

The proposed plant would be built right next door to work hand in hand with the Gilberton Power Company. It will employ gasification and liquefaction, the two processes needed to turn coal into transportation fuel.

Waste anthracite will enter the facility as fine slurry, 65% coal-based carbon and 35% water. This mixture is heated to twenty-five hundred

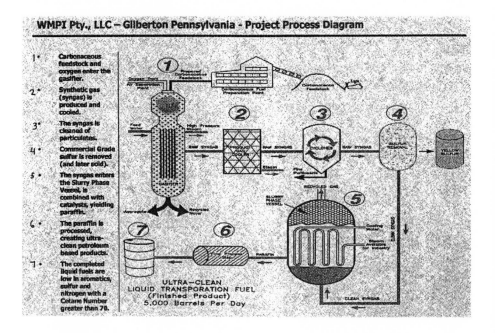

(nn)

degrees Fahrenheit and mixed with oxygen to produce a glass-like material and synthetic gas.

The glass-like byproduct is removed and used in concrete, mortar and plaster. The syngas, meanwhile, continues on to a scrubber that removes fine particulates and sulfur, which is sold to pharmaceutical companies.

Now the clean syngas is combined with catalysts to create a wax-like substance. A finial chemical reaction turns the wax-like material into a liquid form.

A primary benefit of this project is that it applies clean coal technology to address a long-standing environmental reclamation issue associated with the mining and production of coal. This project offers a unique integration of several key technologies to, for the first time; convert forty-seven hundred tons per day of anthracite waste (culm) into forty-one mega-watts of clean electric power and over five-thousand barrels of ultra-clean transportation fuel. This project will process about one million tons per year of coal waste from the Gilberton site. (92)

It has been estimated that, from the past coal mining operations, about two hundred to three hundred million tons of culm can be found across Pennsylvania alone. This technology could be applied in many regions of the country enabling reclamation of lands where coal wastes are stockpiled and significantly reduce waste disposal activities from operating coal mines. The transportation fuel produced will be in the form of ultra-clean high-cetane diesel fuel that will contain no sulfur or aromatics. This fuel can be up-graded to clean burning, reformulated, gasoline. (93)

Today, it is possible for us to turn back to domestic energy sources, and do it cleanly. Coal gasification and coal liquefaction will now allow us to tap the abundant energy stores within our own boarders, without compromising our standards of environmental quality. In fact, these technologies may be our best hope of environmental progress in future years.

Using more of our domestic energy reserves would free us from the reliance on potentially unstable sources and economic drain that results from buying oil from overseas. It would result in greater stability in fuel pricing, job security and enhance our national security by lessening our dependency on foreign sources.

The United States Geological Survey estimates the total coal resources as being 1,600 billion tons.

Currently the United States produces approximately 1,060,000,000 tons of coal annually.

If the United States were to produce from coal, the amount of oil equivalent to what the United States imports, the United States would consume an additional 912 million tons of coal annually.

The United States is sitting on hundreds of years worth of fuel reserves.

To the extent this message becomes clear to our off shore oil suppliers, the perception of a sellers market should diminish as well as the negative implications associated with that scenario and the United States would be positioned to purchase oil on its own terms

In summary, the process utilizes coal waste to produce liquid fuel products that are environmentally friendly known as ultra clean fuels. Fuels meeting these criteria are already required in some areas of the country with strict emissions standards. Ultra clean fuels would not only be plentiful; they could also play a large role in helping us meet the new goals for energy efficiency and clean air.

Making a commitment to ultra clean fuels technology will have substantial and long lasting benefits. Among them:

1. Adoption of ultra clean fuels technology would reenergize the domestic coal production industry.
2. Construction and operation of ultra clean fuels production facilities would create high-quality jobs, improve job security and productivity, and result in numerous spin-off benefits throughout the economy.
3. Reliance on domestic coal resources would revitalize communities in the coal mining producing regions across the country.
4. Lessening dependence on foreign oil sources would improve the United States balance of payments dramatically and reduce the outflow of dollars to overseas suppliers.
5. Diversifying our sources of energy would reduce the threat of war or economic blackmail by foreign powers that control a portion of oil reserves, with potential saving of billions of dollars and thousands of lives.

Some of the environmental benefits of ultra clean fuels are:

1. Ultra clean fuels are cleaner in both production and consumption than standard fossil fuels.
2. Utilizing ultra clean fuels would reduce the overall amount of greenhouse gases introduced into the atmosphere.
3. Ultra clean fuels are generally more environmentally friendly for the production of electricity for electric non-polluting cars.
4. Coal wastes that have blighted the landscape of coal producing regions for decades would be utilized for production, resulting in wholesale reclamation of those regions to their former beauty. (94)

Today, if we chose, we have an opportunity to leave behind for our children and our children's children a nation as strong and self-sufficient as our fathers and grandfathers left us. Or we could continue sucking up foreign oil with the hope our suppliers will continue to ship us all we want at reasonable prices. That's a huge gamble!

As Americans', wouldn't you think the profits created from the sale of the commodity used to manufacture our nations' energy needs stay within our own boarders or would we rather see the profits used to line someone's palace with gold in a far off land?

The choice is ours.
COAL CAN BE KING AGAIN!

Acknowledgments

Most of all I would like to thank my family and friends who supported me throughout this project. Also, I would like to thank the following people who gave me advice and encouragement. Lance E. Metz, Historian of the National Canal Museum, Easton, Pa. Steve Varonka and Christine Goldbeck of the Coal Region Book Nook. Desiree Roffers of Ohio State University. James M. Wallace of Kings College, Wilkes-Barre, Pa. Ed Conrad, Hazelton Standard Speaker. Tracy Cahill, consultant. Rachael Marks of the University of St. Francis. Ralph Hickok of Hickok Sports. John W. Rich, Jr. of Waste Management & Processors, Inc. The United Mine Workers of America. In addition, I cannot forget the "old timers" who shared with me their tears and tales of days gone by.

Credits

1. *Reading Eagle*, Nov. 2, 2003
2. *Reading Eagle*, Nov. 9, 2003
3. Dept. of Energy, Energy Information Administration. (eia.doe.gov.)
4. Donald L. Miller & Richard E. Sharpless, *The Kingdom of Coal: Work, Enterprise, and Ethnic Communities in the Mine Fields* (Easton, Pa: Canal History and Technology Press, 1998), 5.
5. Alfred Mathews and Austin N. Hungerford, *History of the Counties of Lehigh and Carbon in the Commonwealth of Pennsylvania* (Philadelphia: Everts and Richards, 1884) Donald L. Miller & Richard E. Sharpless, *The Kingdom of Coal: Work, Enterprise, and Ethnic Communities in the Mine Fields* (Easton, Pa: Canal History and Technology Press, 1998), 2.
6. encarta.msn.com/encyclopedia—Carboniferous period
7. encarta.msn.com/encyclopedia—Plate technology
8. William E. Edmunds and Edwin F Knoppe, *Pennsylvania, Harrisburg: Topographic and Geological Survey, Commonwealth of Pennsylvania.* 1966, 1-12. Donald L. Miller & Richard E. Sharpless, *The Kingdom of Coal: Work, Enterprise, and Ethnic Communities in the Mine Fields* (Easton, Pa: Canal History and Technology Press, 1998), 5.
9. Donald L. Miller & Richard E. Sharpless, *The Kingdom of Coal: Work, Enterprise, and Ethnic Communities in the Mine Field* (Easton, Pa.: Canal History and Technology Press, 1998), 2.
10. *Reading Eagle*, Nov. 2, 2003
11. Harold W. Aurand, *From the Molly Maguire's to the United Mine Workers: The Social Ecology of an Industrial Union, 1869-1897* (Philadelphia: Temple University Press, 1971), 3-8. Delaware and Hudson Company, *The Story of Anthracite* (New York; Delaware and Hudson, 1932), 11. Donald L. Miller & Richard E. Sharpless, *The Kingdom of Coal: Work, Enterprise, and Ethnic Communities in the Mine Fields* (Easton Pa: Canal History and Technology Press, 1998), 2.
12. Reading Anthracite Company, *The Story of Anthracite*
13. George Korson, *Black Rock; Mining Folklore of the Pennsylvania Dutch* (Baltimore: John Hopkins University Press, 1960), 23. Donald L.

Miller & Richard E. Sharpless, *The Kingdom of Coal: Work, Enterprise, and Ethnic Communities in the Mine Fields* (Easton, Pa: Canal History and Technology Press, 1998), 9.

14. Donald L. Miller & Richard E. Sharpless, *The Kingdom of Coal: Work, Enterprise, and Ethnic Communities in the Mine Fields* (Easton, Pa: Canal History and Technology Press, 1998), 8.

15. Korson, 35-36 Donald L. Miller & Richard E. Sharpless, *The Kingdom of Coal: Work, Enterprise, and Ethnic Communities in the Mine Fields* (Easton, Pa: Canal History and Technology Press, 1998), 10.

16. Delaware and Hudson Company, *The Story of Anthracite* (New York: Delaware and Hudson, 1932), 26-27. Donald L. Miller & Richard E, Sharpless, *The Kingdom of Coal: Work, Enterprise, and Ethnic Communities in the Mine Fields* (Easton, Pa: Canal History and Technology Press, 1998), 11.

17. H. Benjamin Powell, *Philadelphia's First Fuel Crisis.* Donald L. Miller & Richard E. Sharpless, *The Kingdom of Coal; Work, Enterprise, and Ethnic Communities in the Mine Fields* (Easton Pa: Canal History and Technology Press, 1998), 12.

18. Donald L. Miller & Richard E. Sharpless, *The Kingdom of Coal: Work, Enterprise, and Ethnic Communities in the Mine Fields* (Easton, Pa: Canal History and Technology Press), 13.

19. Powell, *Philadelphia's First Fuel Crisis,* chapter 3. Fredrick Moore Binder, *Coal Age Empire: Pennsylvania Coal and its Utilization to 1860* (Harrisburg; Pennsylvania Historical and Museum Commission, 1974), 6. Donald L. Miller & Richard E. Sharpless, *The Kingdom of Coal: Work, Enterprise, and Ethnic Communities in the Mine Fields* (Easton, Pa: Canal History and Technology Press, 1998), 15-16.

20. Reading Anthracite Company, *The Story of Anthracite*

21. Donald L. Miller & Richard E. Sharpless, *The Kingdom of Coal: Work, Enterprise, and Ethnic Communities in the Coal Fields* (Easton, Pa: Canal History and Technology Press, 1998), 47.

22. Korson, 116. Donald L. Miller & Richard E. Sharpless, *The Kingdom of Coal: Work, Enterprise, and Ethnic Communities in the Coal Fields* (Easton Pa: Canal History and Technology Press, 1998), 39-40.

23. Korson, 115. Donald L. Miller & Richard E. Sharpless, *The Kingdom of Coal: Work, Enterprise, and Ethnic Communities in the Mine Fields* (Easton, Pa: Canal History and Technology Press, 1998), 43.

24. Donald L. Miller & Richard E. Sharpless, *The Kingdom of Coal: Work, Enterprise, and Ethnic Communities in the Mine Fields* (Easton, Pa: Canal History and Technology Press, 1998), 44.

25. Donald L. Miller & Richard E. Sharpless, *The Kingdom of Coal: Work, Enterprise, and Ethnic Communities in the Mine Fields* (Easton, Pa: Canal History and Technology Press, 1998), 50.

26. George Korson, *Black Rock: Mining Folklore of the Pennsylvania Dutch* (Baltimore; John Hopkins University Press, 1960), 123. Donald L. Miller & Richard E. Sharpless, *The Kingdom of Coal: Work, Enterprise, and Ethnic Communities in the Mine Fields* (Easton, Pa: Canal History and Technology Press, 1998), 52.
27. Jules I. Bogen, *The Anthracite Railroads: A Study in American Railroad Enterprise* (New York; Ronald Press, 1927), 12-13. Donald L. Miller & Richard E. Sharpless, *The Kingdom of Coal: Work, Enterprise, and Ethnic Communities in the Mine Fields* (Easton, Pa: Canal History and Technology Press, 1998), 53.
28. Korson, 124-125. Donald L. Miller & Richard E. Sharpless, *The Kingdom of Coal: Work, Enterprise, and Ethnic Communities in the Mine Fields* (Easton, Pa: Canal History and Technology Press, 1998), 55-56.
29. Dee Brown, *Hear That Lonesome Whistle Blow: Railroads in the West* (New York: Holt, Rinehart and Winston, 1977), 21-22. Donald L. Miller & Richard E. Sharpless, *The Kingdom of Coal: Work, Enterprise, and Ethnic Communities in the Mine Fields* (Easton, Pa: Canal History and Technology Press, 1998), 53.
30. Louis Poliniak, *When Coal Was King: Mining Pennsylvania's Anthracite in Picture and Story* (Applied Arts Publishers, Lancaster Pa., 1970), 16-17. (Permission to reproduce portions of book *When Coal was King* given by the publisher, Applied Arts Publishers.)
31. James M. Swank, *History of the Manufacture of Iron in All Ages* (Frankline, N. Y.; Burt, 1892), 362. Donald L. Miller & Richard E. Sharpless, *The Kingdom of Coal: Work, Enterprise, and Ethnic Communities in the Mine Fields* (Easton, Pa: Canal History and Technology Press 1998), 64.
32. Donald L. Miller & Richard E. Sharpless, *The Kingdom of Coal: Work, Enterprise, and Ethnic Communities in the Mine Fields* (Easton, Pa: Canal History and Technology Press, 1998), 67.
33. Reading Downtown Improvement District. *DID* (About Reading)
34. Pa. Dept. Of Environment Protection
35. Richard Ramsey Mead, *An Analysis of the Decline of the Anthracite Industry Since 1921* (Ph.D. diss, University of Pennsylvania, 1935). Donald L. Miller & Richard E. Sharpless, *The Kingdom of Coal: Work, Enterprise, and Ethnic Communities in the Mine Fields* (Easton, Pa: Canal History and Technology Press, 1998), 287.
36. Desiree Roffers, *The Coffin Ships*
37. Washington Post, June 26, 1983
38. Avondale Disaster from *The Coal Mines* by Andrew Roy, 134-137. eHistory at OSU
39. Excerpts from *Growing Up in Coal Country* by Susan Campbell Bartoletti. Text copyright (c) 1996 by Susan Campbell Bartoletti. Reprinted by permission of (Houghton Mifflin Company). All rights reserved.

40. Donald L. Miller & Richard E. Sharpless, *The Kingdom of Coal: Work, Enterprise, and Ethnic Communities in the Mine Fields* (Easton, Pa: Canal History and Technology Press, 1998), 112.

41. Mary Siegel Tyson, *The Miners* (Pine Grove, Pa.; Sweet Arrow Lake Press, 1977), 136-37. Donald L. Miller & Richard E. Sharpless, *The Kingdom of Coal: Work, Enterprise, and Ethnic Communities in the Coal Fields* (Easton, Pa; Canal History and Technology Press, 1998), 113.

42. Donald L. Miller & Richard E. Sharpless, *The Kingdom of Coal: Work, Enterprise, and Ethnic Communities in the Mine Fields* (Easton, Pa: Canal History and Technology Press, 1998), 113.

43. Donald L. Miller & Richard E. Sharpless, *The Kingdom of Coal: Work, Enterprise, and Ethnic Communities in the Mine Fields* (Easton, Pa: Canal History and Technology Press, 1998), 144.

44. Donald L. Miller & Richard E. Sharpless, *The Kingdom of Coal: Work, Enterprise, and Ethnic Communities in the Mine Fields* (Easton, Pa: Canal History and Technology Press, 1998), 145.

45. Donald L. Miller & Richard E. Sharpless, *The Kingdom of Coal: Work, Enterprise, and Ethnic Communities in the Mine Fields* (Easton, Pa: Canal History and Technology Press, 1998), 259.

46. Liefer Magnusson, *Company Housing in the Anthracite Region Of Pennsylvania,* U.S. Bureau of Labor Statistics Monthly Labor Review 10 (May 1920), 185. Donald L. Miller & Richard E. Sharpless, *The Kingdom of Coal, Work, Enterprise, and Ethnic Communities in the Mine Field* (Easton, Pa: Canal History and Technology Press, 1998), 286.

47. Donald L. Miller & Richard E. Sharpless, *The Kingdom of Coal: Work, Enterprise, and Ethnic Communities in the Mine Fields* (Easton, Pa: Canal History and Technology Press, 1998), 259.

48. Donald L. Miller & Richard E. Sharpless, *The Kingdom of Coal: Work, Enterprise and Ethnic Communities in the Mine Fields* (Easton, Pa: Canal History and Technology Press, 1998), 319-324.

49. Donald L. Miller & Richard E. Sharpless, *The Kingdom of Coal: Work, Enterprise, and Ethnic Communities in the Mine Fields* (Easton, Pa: Canal History and technology Press, 1998), 122.

50. Excerpts from *Growing Up in Coal Country* by Susan Campbell Bartoletti. Text copyright (c) 1996 by Susan Campbell Bartoletti. Reprinted by permission of (Houghton Mifflin Company). All rights reserved.

51. Donald L. Miller & Richard E. Sharpless, *The Kingdom of Coal: Work, Enterprise, and Ethnic Communities in the Mine Fields* (Easton, Pa: Canal History and Technology Press, 1998), 122.

52. Donald L. Miller & Richard E. Sharpless, *The Kingdom of Coal: Work, Enterprise and Ethnic Communities in the Coal Fields* (Easton, Pa: Canal History and Technology Press, 1998), 125.

53. Excerpts from *Growing Up in Coal Country* by Susan Campbell Bartoletti. Text copyright (c) 1996 by Susan Campbell Bartoletti. Reprinted by permission of (Houghton Mifflin Company). All rights reserved.

54. Lyrics obtained from the George Korson estate by James Wallace, Kings College, Wilkes-Barre, Pa.

55. Andrew M. Greeley, *That Most Distressful Nation; The Taming of the American Irish* (Chicago: Quadrangle Books, 1972), 28. Donald L. Miller & Richard E. Sharpless, *The Kingdom of Coal: Work, Enterprise, and Ethnic Communities in the Mine Fields* (Easton, Pa: Canal History and Technology Press, 1998), 138.

56. *The Mollies and the A.O.H.*, by Walter Boyle

57. Donald L. Miller & Richard E. Sharpless, *The Kingdom of Coal: Work, Enterprise, and Ethnic Communities in the Mine Fields* (Easton, Pa: Canal History and Technology Press, 1998), 154-155.

58. Donald L. Miller & Richard E. Sharpless, *The Kingdom of Coal: Work, Enterprise, and Ethnic Communities in the Mine Fields* (Easton, Pa: Canal History and Technology Press, 1998), 161-162.

59. Kevin Kenny, *Making Sense of the Molly Maguires* (Oxford University Press, New York, N.Y., 1998), 164-166. Reproduced by permission of Oxford University Press, Inc.

60. Donald L. Miller & Richard E. Sharpless, *The Kingdom of Coal: Work, Enterprise, and Ethnic Communities in the Mine Fields* (Easton, Pa:, Canal History and Technology Press, 1998), 156-158.

61. Kevin Kenny, *Making Sense of the Molly Maguires* (Oxford University Press, New York, N.Y., 1998), 190-192. Reproduced by permission of Oxford University Press, Inc.

62. Joseph H. Bloom, *Undermining the Molly Maguires. American History Magazine*.

63. Donald L. Miller & Richard E. Sharpless, *The Kingdom of Coal: Work, Enterprise, and Ethnic Communities in the Mine Fields* (Easton, Pa., Canal History and Technology Press, 1998), 164-169.

64. *Memory of the Molly Maguires Kept Alive*, By Marigrace Heyer

65. Sidney Lens, *The Labor Wars: From The Molly Maguires to the Sit Downs* (Garden City, N.Y.; Anchor, Doubleday, 1974), 11-13. Donald L. Miller & Richard E. Sharpless, *The Kingdom of Coal: Work, Enterprise, and Ethnic Communities in the Mine Fields* (Easton, Pa., Canal History and Technology Press, 1998), 136.

66. Joseph H. Bloom, *Undermining the Molly Maguire's. American History Magazine.*

67. Michael Doyle, (Mauch Chunk Gazette, July 14, 1877)

68 Donald L. Miller & Richard E. Sharpless, *The Kingdom of Coal: Work, Enterprise, and Ethnic Communities in the Mine Fields* (Easton, Pa., Canal History and Technology Press, 1998), 149.

69. Sidney Lens, *The Labor Wars: From the Molly Maguires to the Sit Downs* (Garden City, N.Y. Doubleday, 1974), 18-19. Donald L. Miller & Richard E. Sharpless, *The Kingdom of Coal: Work, Enterprise, and Ethnic Communities in the Mine Fields* (Easton, Pa., Canal History and Technology Press, 1998), 151.

70. Donald L. Miller & Richard E. Sharpless, *The Kingdom of Coal: Work, Enterprise, and Ethnic Communities in the Mine Fields* (Easton, Pa., Canal History and Technology Press, 1998), 123-124.

71. *A Pictorial Walk Through the 20th Century, The Irish in Mining*, (Dept. of Labor, Mine Safety and Health Administration).

72. Kenneth Wolensky, *Lattimer Massacre* (Pennsylvania Historical and Museum Commission).

73. Donald L. Miller & Richard E. Sharpless, *The Kingdom of Coal: Work, Enterprise, and Ethnic Communities in the Mine Fields* (Easton, Pa., Canal History and Technology Press, 1998), 247.

74. Donald L. Miller & Richard E. Sharpless, *The Kingdom of Coal: Work, Enterprise, and Ethnic Communities in the Mine Fields* (Easton, Pa., Canal History and Technology Press, 1998), 249.

75. UMWA, *A Brief History of the United Mine Workers of America.*

76. Rachael Marks, *Anthracite Coal Strike of 1902* (1999)

77. Donald L. Miller & Richard E. Sharpless, *The Kingdom of Coal: Work, Enterprise, and Ethnic Communities in the Mine Fields* (Easton, Pa., Canal History and Technology Press, 1998), 246.

78. Donald L. Miller & Richard E. Sharpless, *The Kingdom of Coal: Work, Enterprise, and Ethnic Communities in the Mine Fields* (Easton, Pa., Canal History and Technology Press, 1998), 255.

79. UMWA, *A Brief History of the United Mine Workers of America.*

80. Story compiled from my recollection of the event, combined with various articles from the *Pottsville Republican* and *Hazelton Standard-Speaker* from the last two weeks of August 1963.

81. *Life after Death*, by Ed Conrad

82. Story compiled from my recollection of the event, combined with various articles from the *Pottsville Republican* and *Hazelton Standard Speaker* from the last two weeks of August 1963.

83. Donald L. Miller & Richard E. Sharpless, *The Kingdom of Coal: Work, Enterprise, and Ethnic Communities in the Mine Fields* (Easton, Pa: Canal History and Technology Press, 1998), 147-149.

84. Donald L. Miller & Richard E. Sharpless, *The Kingdom of Coal: Work, Enterprise, and Ethnic Communities in the Mine Fields* (Easton, Pa: Canal History and Technology Press, 1998), 183.

85. Donald L. Miller & Richard E. Sharpless, *The Kingdom of Coal: Work, Enterprise, and Ethnic Communities in the Mine Fields* (Easton, Pa: Canal History and Technology Press, 1998), 203.

86. Donald L. Miller & Richard E. Sharpless, *The Kingdom of Coal: Work, Enterprise, and Ethnic Communities in the Mine Fields* (Easton, Pa: Canal History and Technology Press, 1998), 199.
87. Pottsville Maroons, 1925-28. Boston Bulldogs 1929, Hickok Sports
89. Pottsville Maroons, 1925-28. Boston Bulldogs 1929, Hickok Sports
90. Pete Gray, *Historic Baseball*
91. Ultracleanfuels, WMPI, LLC
92. Department of Energy
93. Department of Energy
94. Ultracleanfuels, WMPI, LLC

Pictures

(Cover) Gerald L. McKerns

a.   Angel, Gerald L. McKerns
      A nineteenth-century miner, Courtesy of the National Archives

b.   Gerald L. McKerns

e.   Courtesy of the Pennsylvania Historical and Museum Commission.

f.   Courtesy of the Pennsylvania Canal Society Collection, Canal Museum, Easton, Pa.

g.   berksweb.com/histsoc/canal/images

h.   Courtesy of the Historical Society of Schuylkill County, Pottsville, Pa.

i.   Courtesy of the Historical Society of Schuylkill County, Pottsville, Pa.

j.   Courtesy of the Wyoming Historical and Geological Society, Wilkes-Barre, Pa.

k.   Courtesy of the Burg Collection—m273—Pennsylvania State Archives.

l.   undergroundminers.com/middle field

m.   Courtesy of the George Bretz Collection, Albin O. Kuhn Library, University of Maryland, Baltimore County

n.   Louis Poliniak, *When Coal was King* (Applied Arts Publishers, Lancaster, Pa: 1996), 13. (Permission to reproduce portions of book *When Coal was King* given by the publisher, Applied Arts Publishers.)

o.   Louis Poliniak, *When Coal was King* (Applied Arts Publishers, Lancaster, Pa: 1996), 13. (Permission to reproduce portions of book *When Coal was King* given by the publisher, Applied Arts Publishers.)

p.   Courtesy of the Lewis Hine Collection, Albin O Kuhn Library, University of Maryland, Baltimore County

q.   Courtesy of the Lewis Hine Collection, Albin O. Kuhn Library, University of Maryland, Baltimore County

r.   Louis Poliniak, *When Coal was King* (Applied Arts Publishers, Lancaster, Pa. 1996), 21. (Permission to reproduce portions of book *When Coal was King* given by the publisher, Applied Arts Publishers.)

v.   Courtesy of the Historical Society of Schuylkill County, Pottsville Pa.

w.   Courtesy of the Historical Society of Schuylkill County, Pottsville Pa.

y.   Courtesy of the Library of Congress

z.   Courtesy of the Historical Society of Schuylkill County, Pottsville Pa.

aa. Collection of Daniel Pascavage.
bb. Courtesy of the Historical Society of Schuylkill County, Pottsville Pa.
cc. Courtesy of the Historical Society of Schuylkill County, Pottsville Pa.
dd. Alexander Campbell Sketch by Arron Borger. News Paper sketch's from *The Daily Graphic*, New York and reprinted in the *The Old Jail Museum and Molly Maguires* publication, Jim Thorpe, Pa.
ee. Courtesy of the Library of Congress
ff. Courtesy of the Burg Collection, Pennsylvania State Archives
gg. Courtesy of the Burg Collection, Pennsylvania State Archives
hh. Courtesy of the Library of Congress
ii. Courtesy of the Library of Congress
jj. Gerald McKerns
kk. Courtesy of Mary Pompei
ll. Courtesy of the Wyoming Historical and Geological Society, Wilkes-Barre, Pa.
mm. Pete Gray, *Famous Coal Crackers*
nn. WMPI Pty, llc, Gilberton, Pa.

Bio. 1

Gerald McKerns was born in a company owned house in a coal mining "patch" named St. Nicholas located in the heart of the Pennsylvania's anthracite "coal region". Upon completing High School Gerald migrated to Reading, Pennsylvania were he spent nearly thirty years working in a steel mill. Although living out of the area for many years Gerald never forgot his roots. You can take the coal cracker out of the "coal region" but you can't take the "coal region" out of a coal cracker.

Bio. 2

Gerald McKerns was born in a company owned house in a coal mining patch named St. Nicholas located in the heart of Pennsylvania's hard "coal region" This small village sat in the shadow of the largest coal breaker in the world. While attending school Gerald's focus was geared more towards the day he could get a job in the coal industry rather than on his schoolbooks. School didn't seem important.

During this time, the nation was making a transition from coal to gas and oil for home heating which resulted in the demand for hard coal to dwindle. The anthracite coal industry was depressed. Hard times hit the area. The jobs were gone.

Like so many before and so many to follow, Gerald had to leave the area to find a brighter future. His exodus took him to a steel mill in Reading, Pennsylvania.

Over the years, in the mill, Gerald found himself often bending the ears of his co-workers talking about the coal region. On many occasions, he would hear the remark, "You should write a book", to which he'd reply, "I can't even talk right and you want me to write a book?"

When Gerald left the mill in 2002 he found himself with some idle time on his hands. The echo kept ringing in his ears, "You should write a book." Listening to that echo, the project began. Gerald hoped that writing on the subject that was close to his heart would make up for his lack of literary skills. It was quite a challenge. But one word at a time, one page at a time, one chapter at a time, and with many set backs the project was completed. The title of his book is *The Black Rock that Built America.* By Gerald McKerns CC.

Gerald explains the "CC." following his name this way. "I graduated from high school near the bottom of my class. After nearly thirty years of sweat and blood in a steel mill, I now realize the importance of an education. But the education I got out of the class room, growing up in the coal region, could not be obtained from any books.

I watched men get old before their time, their lungs blackened with coal dust from the years of working down below. The only bright spot in their day was a plug of chewing tobacco and a mug of beer.

I watched wives and children waiting out side a caved in mine, clinging to one another for support, hoping the rescue workers would bring their loved ones out alive. Sometimes they did and sometimes they didn't.

I listened to the tales the old timers told, some with laughter and some with tears.

There isn't any book that could teach you to appreciate life the way you learned it in the hard coal region of Pennsylvania." Gerald McKerns CC (Coal Cracker).

Roots grow deep in the hard coal region.

You can take a coal cracker out of the coal region but you can't take the coal region out of a coal cracker.

Book Summary 1

The Black Rock That Built America explains how, on the backs of thousands of European immigrants, America was transformed from a mostly rural nation into the world's greatest industrial power. As the nation expanded in the nineteenth century, anthracite coal fueled the making of steel, the building of railroads, the operation of factories, and the heating of homes. This book tells of the struggles these immigrant miners endured while performing the grueling and dangerous work of extracting anthracite coal from the earth in order to earn their place in America.

Book summary 2

This country, as we know it today with its economic prosperity and military might, didn't just happen. It took years of hard work by the many that came before us. We, as a nation, are blessed with vast natural resources. But our greatest natural resource is our people. The people who took a wilderness and with their sweat and blood would have it evolve into what we have today. With our constitution as its cornerstone and our people as its building blocks, a great nation did emerge.

Many of our children aren't aware of the hardships their grandfathers and great-grandfathers endured to sculpt this nation into its grandeur.

Every day has a yesterday and every today has a tomorrow. As each tomorrow arrives, each yesterday gets further into the past. The stories our great-grandfathers handed down, by word of mouth get fewer with each passing generation and some day they will be no more.

To appreciate where we are today it is essential that we know where we came from.

The Black Rock that Built America explains how important an area, consisting of only about five-hundred square miles in northeastern Pennsylvania, was in propelling this vast wilderness into a global giant.

For over one-hundred years, the hard coal region was the starting point that led America down the road to being the world leader.

The anthracite coal region of Pennsylvania and its people should get their rightful place in history. The lives of these immigrant miners should be celebrated. This book will help in that cause.

It all began in the early nineteenth century for those looking to escape the poverty and unemployment of Europe. The anthracite region of northeastern Pennsylvania appeared to be the land of opportunity. They would come by the thousands, drawn by the promise of steady work in the mines. At one point, twenty different languages were spoken in the hard coal region. Most of the immigrants were hired by the mines and did the grueling and dangerous work of extracting coal from the earth in order to find their place in America.

The wages were low but the price was high. Over thirty thousand men and boys died while mining the anthracite coal in the mountains of northeastern Pennsylvania.

Jobs in the mines were plentiful. At first, these jobs seemed like a blessing to the destitute immigrants. The coal companies provided company owned housing with rent deducted from paychecks and company owned stores, where a bill could be accumulated. But after the house rent and excessive prices at the company store were deducted from their paychecks, there was very little left. They were trapped in a system that promised them only generations of survival at the poverty level. Their young children would have to work in the breakers for the families to survive and woe to the family whose father was killed or crippled in the mines. Without an income to pay for the housing, the family was turned out into the muddy streets to beg or rely on friend's charity.

The one unifying experience in the lives of the miners was their work. Mining was the most dangerous job of the day. To the hazards underground were added exploitations of other kinds. Miners were paid for the coal they dug. Unscrupulous company employees "short weighed" miners production. The miners paid for their blasting powder, tools, and other supplies they used. The operators forced the miners to buy their supplies from the company store as a condition of employment.

The miner became the slave of the coal barons. The miner had to fight back in form of unionization. This resulted in more men dying. Some were hung and some were shot.

Not only did the miner have to stand up to the mine owner but also they had to face the dangers of working down below. Included in the list of thousands of miners who lost their lives in the anthracite coal industry were many boys, some as young as ten years of age.

The cycle of a miner's life began early, sometimes as young as six years old. In the deep snows of winter, fathers carried smaller boys to the breaker on their backs in the predawn darkness, or mothers would take the younger boys and return to wait for them at the end of their shift.

So many of these young lads went from the cradle to the grave with not much in between and the odds that some of these boys would reach their senior years were not very good. But even if they did, there were no pensions, no health care, no social security, and no nest egg. Their later years would be as dismal as the years before.

As Americans, we have become accustomed to a lifestyle that is unparalleled anywhere else in the world. Sometimes we take it for granted. One must remember, that today, we are the beneficiaries of the sacrifice of those miners who were willing to put their lives on the line in order that their children and their children's-children and now our children, could have a better existence.

Although many of these coal crackers didn't have much schooling, some of the lessons learned from their living experiences in and around the mines couldn't be taught in any books. They learned the value of life. Each day they would enter the mines not knowing if they would return home to their loved ones. If by fate, the *whistle* didn't blow and the men and boys returned home safely it would be cause for jubilation. They learned to celebrate life. Be it at work or play they lived life to the max. Because just maybe tomorrow the *whistle* would blow and a wife would become a widow and a child would be without a father.

The kingdom of coal is gone. The black rock that broke America's dependency on foreign coal, fueled an industrial revolution, kept millions warm, created great wealth, and birth to a vibrant immigrant culture has severed its time in history. Anthracite's final legacy is a warning to all Americans that human lives and natural resources are finite and precious, they can no longer be sacrificed indiscriminately on the alter of private greed.

The anthracite coal miners of northeastern Pennsylvania contributed much to this great nation. Let us not forget them.